WITHDRAWN
FROM STOCK

CANCELLED

A New Look at
MECHANISMS IN BIOENERGETICS

A New Look at MECHANISMS IN BIOENERGETICS

Efraim Racker
Division of Biological Sciences
Cornell University
Ithaca, New York

ACADEMIC PRESS New York San Francisco London 1976

A Subsidiary of Harcourt Brace Jovanovich, Publishers

Cover illustration by Efraim Racker

ACADEMIC PRESS, INC.
111 Fifth Avenue, New York, New York 10003

United Kingdom Edition published by
ACADEMIC PRESS, INC. (LONDON) LTD.
24/28 Oval Road, London NW1

Library of Congress Cataloging in Publication Data

Racker, Efraim, Date
A new look at mechanisms in bioenergetics.

"Based on the Robbins lectures given at Pomona College in April 1973."
Bibliography: p.
Includes index.
1. Bioenergetics. 2. Oxidation, Physiological. 3. Phosphorylation. I. Title.
QH510.R3 574.1'9121 75-44763
ISBN 0-12-574670-9 (cloth)
ISBN 0-12-574672-5 (paper)

PRINTED IN THE UNITED STATES OF AMERICA

Contents

Preface

> You say you are scarcely competent to write books just yet. That is just why I recommend you to learn. If I advised you to learn to skate, you would not reply that your balance was scarcely good enough yet. A man learns to skate by staggering about making a fool of himself. Indeed he progresses in all things by resolutely making a fool of himself.
>
> **George Bernard Shaw**
> *Advice to a Young Critic*

This book is based on the Robbins Lectures given at Pomona College in April, 1973. Since I had written them before I delivered them to the enthusiastic students of this College, I did not hesitate to agree to have them published after appropriate revisions. However, during the past years so much has happened in the field of bioenergetics that whenever I finished making corrections in the last lectures, I discovered that the early lectures needed rewriting. When I finished those, the last lectures were out of date.

Although I learned to skate at the early age of five, I did not start writing a book until I was past fifty, and I never quite learned the proper balance. Therefore I "resolutely" take refuge in Shaw's advice. I shall not dazzle the readers with mental pirouettes and I shall be glad if I can glide on the icy surface without mishap.

This book is not a new edition of my previous book on "Mechanisms in Bioenergetics." I have avoided repetition and, instead, have frequently referred to the other book. Although this occasionally may have resulted in some discontinuity of thought, it had the advantage of keeping the size of this book small.

As in the first book, the lectures are based mainly on work performed in my laboratory. Although I frequently cite important publications of other investigators, I probably have failed to give credit to some significant contributions. These omissions are in most cases unintentional and are only an index of my inability to cope with the ever increasing and staggering literature in the field of bioenergetics. In a few instances I have intentionally refrained from discussing some controversial experiments or hypotheses of oxidative phosphorylation which I believe contribute at present little to the experimental approach. These omissions will serve as an index of my poor judgment. When I omitted in my first book a discussion of the chemiosmotic hypothesis of Mitchell, it was only because I failed to recognize its potential as a working hypothesis. I believe I have adequately made up for this defect in the current book. Perhaps scientists are much too concerned with giving and receiving credits. The research work in biochemistry in the twentieth century is probably like the building of cathedrals in the middle ages. It is the work of many, and the identity of those who participated in their creation is a matter of little consequence and will soon be forgotten.

I am addressing these lectures to the young students of biology and biochemistry who want to devote their lives to research. At present the spirits of the scientific community are not very high. Society, including legislative bodies of many countries, has become less sympathetic to basic research, which is regarded as a luxury. Fluctuation in sympathy of society to science is not new. When Faraday demonstrated his first experiments on electricity, we are told that Gladstone, who later became the Chancellor of the Exchequer, asked what it was good for. Faraday replied, "One day, sir, you may tax it." I tell this story not only to show that Gladstone was not interested in basic science, but also that Faraday's imagination reached out to a future that was far away from reality. It would be disastrous if we guided our efforts in basic science by consideration of their usefulness. But I see nothing wrong with using our collective imagination to exploit the usefulness of basic discoveries once they have been made. Perhaps some of our relevance-conscious students will help to establish better channels of communication between the basic and applied scientist of the next generation.

I acknowledge with gratitude the financial support I have received over the years from the National Cancer Institute, the National Science Foundation, and the American Cancer Society. The work on which this book is based would not have been possible without this support and the collaboration of colleagues and of many young and gifted students and postdoctoral fellows who have spent two or more years in my laboratory. I should like to mention those who have greatly contributed to the more recent experimental work discussed in this book: W. Arion, R. Berzborn, A. Bruni, B. Bulos, C. Burstein, C. Carmeli, R. Carroll, R. Christiansen, D. Deters, E. Eytan, G. Eytan, J. Fessenden-Raden, L. Fisher, G. Hauska, P. Hinkle, Y. Kagawa, A. Kandrach, B. Kanner, A. Knowles, E. LaBelle, D. Lang, S. Lien, A. Loyter, G. McCoy, C. Miller, N. Nelson, H. Nishibayashi, I. Ragan, D. Schneider, P. Scholnick, R. Serrano, H. Shertzer, E-M. Suolinna, and J. Telford. Special thanks are due to Mr. Mike Kandrach, who keeps our laboratories running, and to Mrs. Judy Caveney, my secretary, whose devotion, patience, and editorial assistance were essential for the completion of this book. I am grateful for the comments of G. Schatz, G. Eytan, and C. Miller during the preparation of the manuscript. Last, but not least, I want to extend my thanks to my wife Franziska and our daughter Ann for claiming that they had not noticed that I was writing this book.

Efraim Racker

Abbreviations

A, B	Members of the oxidation chain
A-particles	Submitochondrial particles prepared by sonication of bovine heart mitochondria in the presence of ammonia and EDTA
AMP, ADP, ATP	Adenosine 5′-mono, di-, and triphosphate
ANS	1-Anilino-8-naphthalene sulfonate
AS-particles	A-particles passed through Sephadex
ASU-particles	A-particles passed through Sephadex and treated with urea
ATPase	Adenosine triphosphatase
BHK	Baby hamster kidney cells
C-side	The side of the inner mitochondrial membrane which faces the outer mitochondrial membrane and contains cytochrome *c*
$CF_0 = F_0$	A membranous preparation from mitochondria conferring oligomycin (or rutamycin) sensitivity to F_1
CF_1	Coupling factor 1 from chloroplasts
COV	Reconstituted liposomes, containing cytochrome oxidase as the only protein
DABS	Diazobenzene sulfonate
DBMIB	2,5-Dibromo-3-methyl-6-isopropyl-*p*-benzoquinone-dibromothymoquinone
DCCD	*N,N′*-Dicyclohexylcarbodiimide
DCMU	Dichlorophenyl-1,1-dimethyl urea
DNP	2,4-Dinitrophenol
DPN = NAD	Nicotinamide adenine dinucleotide
EDAC	1-Ethyl-3(3-dimethylaminopropyl) carbodiimide
EDTA	Ethylenediaminetetraacetic acid
EGTA	Ethyleneglycol-bis-*N,N′*-tetraacetic acid

ETP_H	Unresolved submitochondrial particles yielding high P:O ratios
Factor A, B	Coupling factors of oxidative phosphorylation resembling the properties of F_1 and F_2
FCCP	Carbonylcyanide *p*-trifluoromethoxyphenylhydrazone
Fd	Ferredoxin
Fp	Flavoprotein
F_1, F_2, F_3, F_4, F_5, F_6	Coupling factors 1 (ATPase) 2,3,4,5, and 6, respectively
HP	Hydrophobic protein, another name for F_0
M-side	The side of the inner mitochondrial membrane which faces the matrix
NBD-chloride	7-Chloro-4-nitrobenzo-2-oxa-1,3-diazole
NEM	*N*-Ethylmaleimide
NHI protein	Nonheme iron protein
OSCP	Oligomycin sensitivity conferring protein
PC	Phosphatidylcholine
PCB	Phenyl dicarbaundecaboron
PE	Phosphatidylethanolamine
P_i	Inorganic orthophosphate
PEP	Phospoenolpyruvate
PK	Pyruvate kinase
PMS	*N*-Methyl phenazonium methosulfate
PQ	Plastoquinone
Q	Coenzyme Q or ubiquinone
RCR	Respiratory control ratio, expressed as oxidation rate in the presence of uncoupler/in the absence of uncoupler
SDS	Sodium dodecyl sulfate
SMP	Submitochondrial particles prepared by sonication of bovine heart mitochondria in the presence of pyrophosphate
STA-particles	A-particles treated with silicotungstate
S_{13}	5-Chloro-3-*tert*-butyl 2′-chloro-4′-nitrosalicylamalide
TPN = NADP	Nicotinamide adenine dinucleotide phosphate
Tris	Tris(hydroxymethyl)aminomethane
TU-particles	Submitochondrial particles prepared by stepwise exposure of light layer submitochondrial particles to trypsin and urea
TUA-particles	TU-particles exposed to sonic oscillation in the presence of dilute ammonia (pH 10.4)
X, Y	Members of the coupling device
1799	Bis(hexafluoroacetonyl)acetone
3T3	Mouse embryo fibroblasts

Lecture 1

Troubles Are Good for You

> If you have built a perfect demonstration do not remove all traces of the scaffolding by which you have raised it.
>
> **Clark Maxwell**

How It All Started

Before going into the complex details of the structure and function of mitochondrial and chloroplasts membranes, I would like to transmit to you some of the general lessons I have learned in doing research in the field of bioenergetics. When you read excellent scientific articles in popular journals such as *Scientific American* or even in professional journals, you will be much impressed. You are exposed to lucid expositions, sometimes brilliant experiments, important developments, and visions of an even more interesting future. The data are usually unambiguous and the conclusion convincing. The writer may even succeed in transmitting some of the excitement of the laboratory and you appreciate it. But almost invariably certain aspects of the work will be missing. How did the discovery really come about? How much sweat and trouble was there in its making? How much of it was thought out beforehand; how much was accidental? Since these lectures are primarily addressed to students I would like to transmit, at least in the first one, a picture of research life which I believe is somewhat closer to reality. It contains large shares of troubles, doubts, serendipity, and, last but not least, interaction between different investigators. I believe that we should prepare our students for these aspects, not only to avoid disappointment, but to convey to them the concept that troubles and doubts are seeds of the future. They can lead us, if we follow them, into unknown territory and challenging problems. To paraphrase Goethe with equal exaggeration: For the true scientist nothing is more difficult to bear (and indeed suspect) than an uninterrupted series of beautiful experiments. Or as Piet Hein says in a grook: "Problems worthy of attack prove their worth by hitting back."

I want to give you first a personal account of how I got to the problem of oxidative phosphorylation and to tell you some of the things that happened on the way to the palace or to the castle. You have a choice here depending on whether you consider my account a fairy tale or a Kafka nightmare.

Let us start out with the first basic question. How does a biochemist choose a problem to work on? More specifically, what is a nice M.D. doing in oxidative phosphorylation rather than working on a relevant problem in medicine?

When I was a medical student almost 40 years ago, I wanted to be a psychiatrist, I wanted to understand mental diseases, I wanted to cure and heal the psychotic mind. This was a relevant problem, pregnant with economical and social implications. Having been raised in Vienna and put to sleep by the lullably of the Oedipus complex, I first turned toward the teaching of Freud, but could not find a firm footing. I was soon plagued by doubts fortified by a statement of Freud that psychosis is a child of the night, and cannot be cured through the mind. In fact, he believed in the organic genesis of psychosis.

In 1938 when a mass psychosis invaded Vienna I left for England and joined the laboratory of Dr. Quastel who had written a fascinating article on the relationship of mental disorders and amines (Quastel, 1936). I became interested in mescaline, benzedrine (better known as speed) and other analogues of biogenic amines, which in large doses produce illnesses resembling psychoses. Thus I became in 1938 a biochemical hippy. Coming to the United States in 1941, I found no interest in biogenic amines, but there were funds for studies in applied research in the field of poliomyelitis. My research was supported by the "March

TABLE 1-1 **Glyceraldehyde-3-phosphate Dehydrogenase**

Mechanism I

1. $R{-}C(=O)H + H_3PO_4 \rightleftharpoons R{-}C(OPO_3H_2)(OH)H$

2. $R{-}C(OPO_3H_2)(OH)H + DPN^+ \rightleftharpoons R{-}C(=O)OPO_3H_2 + DPNH + H^+$

Mechanism II

1. $R{-}C(=O)H$ + Enzyme SH + $DPN^+ \rightleftharpoons$ Enzyme-S$-C(=O)-R + DPNH + H^+$

2. Enzyme-S$-C(=O)-R + H_3PO_4 \rightleftharpoons R{-}C(=O)OPO_3H_2$ + Enzyme SH

of Dimes" and I had the impressive salary of 12,000 dimes per year. In exploring the effect of polioviruses on brain metabolism I observed a defect in glycolysis which I traced to an inactivation of glyceraldehyde-3-phosphate dehydrogenase (Racker and Krimsky, 1948). In examing this key enzyme of glycolysis more closely I discovered that the popular hypothesis of its mode of action was incorrect. Warburg and Christian (1939) had proposed a simple and ingenious mechanism based on a chemical model system shown in Mechanism I of Table 1-1. The first step is the formation of an adduct between the aldehyde and inorganic phosphate which is directly oxidized to 1,3-diphosphoglycerate. Warburg's influence on the biochemical society was so great that his formulation was not only blindly accepted in all textbooks but an enterprising firm in New York City sold the hypothetical adduct diphosphoglyceraldehyde for about $1000 per 100 mg. Nevertheless, we could show without ambiguity (Racker and Krimsky, 1952) that the enzyme catalyzes the two steps shown in Mechanism II, Table 1-1. The first step is an oxidation of the aldehyde–enzyme complex to an acyl enzyme intermediate which we obtained in crystalline form (Krimsky and Racker, 1955). Phosphate enters in a second step by cleaving the thiol ester of the enzyme yielding 1,3-diphosphoglycerate. Phosphoglycerate kinase transfers the phosphate in the 1-position of this compound to ADP yielding ATP. Warburg never accepted this formulation since his chemical model was simpler and better. Let us look at the first two lessons we can learn from this account.

Lesson 1: Chemistry is good, nature is better.

Lesson 2: If there is no money available for fundamental research start working on a project of applied research. If you proceed logically you will soon be doing basic research.

You can see now how I became interested in oxidative phosphorylation. My turning to these fundamental problems was not caused by a loss of interest in biogenic amines, but by a clear realization that without the fundamentals of biochemistry we cannot understand either physical or mental disorders. Actually I have started recently to work again with biogenic amines (McCauley and Racker, 1973). Having fled in my youth from a mass pychosis in Germany, facing at present a drugged society blinded by acute pains, I cannot think of a more relevant problem to work on.

What Is Oxidative Phosphorylation?

Let us look at the processes that take place in the inner mitochondrial membrane. As shown in Fig. 1-1 the basic reaction is oxidation. Substrates such as pyruvate enter the Krebs cycle and donate hydrogens to DPN. The reduced nucleotide (DPNH) is oxidized by the mitochondrial oxidation chain in discrete steps which permit conservation of energy of oxidation and formation of ATP. In fact, the formation of ATP is compulsorily coupled to the oxidation process so that respiration ceases when no ADP and P_i are available. We shall return to this important regulatory mechanism called "respiratory control." There are three classes of compounds which interfere with oxidative phosphorylation.

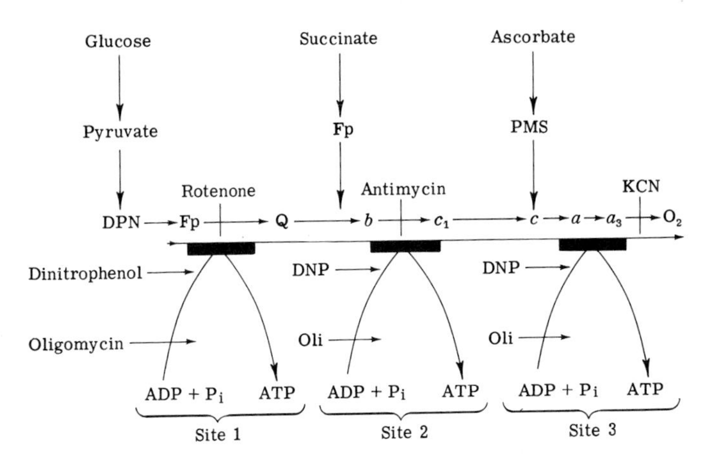

Fig. 1-1 Oxidative phosphorylation in mitochondria.

There are (1) inhibitors of the oxidation chain such as cyanide or antimycin; (2) uncouplers such as 2,4-dinitrophenol or carbonylcyanide *p*-trifluoromethoxyphenylhydrazone which abolish the conservation of energy and give rise to a dissipation of the oxidation energy into heat; and (3) energy transfer inhibitors such as oligomycin or rutamycin (Lardy *et al.*, 1958) which prevent the conversion of the oxidation energy into ATP. These energy transfer inhibitors block the formation of ATP coupled to oxidation as well as the breakdown of ATP by mitochondrial ATPase. As will be elaborated later, the oligomycin-sensitive ATPase activity represents a partial reversal of the process of oxidative phosphorylation.

There are three ATP molecules generated for each DPNH which is oxidized by molecular oxygen and we speak of a P:O ratio of 3, an expression of the efficiency of the process. It was shown many years ago (Ochoa, 1943) that for each molecule of pyruvate oxidized there are 15 molecules of ATP formed from ADP and inorganic phosphate. However, pyruvate has only 4 hydrogens to donate to 2 oxygens and with a P:O ratio of 3 should therefore yield only 6 molecules of ATP (Table 1-2).

TABLE 1-2 **Where Are the Hydrogens Coming From?**

	Oxidation		ATP
1. Pyruvate + 2.5 O_2 → 2 H_2O + 3 CO_2 ($CH_3COCOOH$)	4 H	→	6
2. Pyruvate + 3 H_2O + 2.5 O_2 → 5 H_2O + 3 CO_2	10 H	→	15

Where are the other 6 hydrogens coming from so that 15 ATP can be formed? The answer to this question gives us what I think is the key to the puzzle why nature has designed the complex acrobatic scheme of the Krebs cycle. Its major purpose I believe is to increase the energy yield by catalyzing the cleavage of water. There are three steps at which water enters the Krebs cycle—one at the transformation of fumarate to malate and the other two somewhat more indirectly during the utilization of acetyl-CoA and succinyl-CoA. In the course of the Krebs cycle the hydrogens of these water molecules are separated from the oxygen and are donated to DPN or to a flavoprotein (e.g., succinate dehydrogenase) and then transported via the oxidation chain of mitochondria as electrons and protons as we shall discuss later.

I want to tell you of an incident which happened a few years after Ochoa had established by ingenious and simple experiments that the P:O ratio is 3 (Ochoa, 1943). One of the leading physical chemists in England published a long review with extensive thermodynamic calculations stating that the theoretic maximal P:O ratio is 2 (Ogston and Smithies, 1948). I remember that I was quite upset when this article appeared and brought it to one of our daily luncheon discussions with Dr. Ochoa. But Dr. Ochoa was amused and shrugged the criticism off; he said "time will tell." Indeed it did. The mathematical calculations were all correct but were based on an erroneous value for the free energy of hydrolysis of ATP. We may stop here for the next lesson.

> *Lesson 3:* A clean experiment is worth more than a few hundred dirty calculations.

How Do We Measure Oxidative Phosphorylation?

As illustrated in Fig. 1-2 there are several methods available for measuring oxidative phosphorylation. The older methods of manometric determination of oxygen uptake has been replaced by the polarographic method. Disappearance of P_i, the classical method used by Ochoa, has been replaced by the more sensitive and accurate method with $^{32}P_i$, which is incorporated via ATP into glucose 6-phosphate in the presence of an excess of hexokinase. This procedure has the advantage that it is applicable to all systems catalyzing oxidative phosphorylation even in the presence of an active ATPase. The latter becomes a serious problem with submitochondrial particles which hydrolyze ATP quite rapidly. The polarographic procedure of Chance and Williams (1956) of measuring oxidative phosphorylation depends on the stimulation of respiration by addition of small amounts of ADP. It is simple and very suitable for intact mitochondria which have

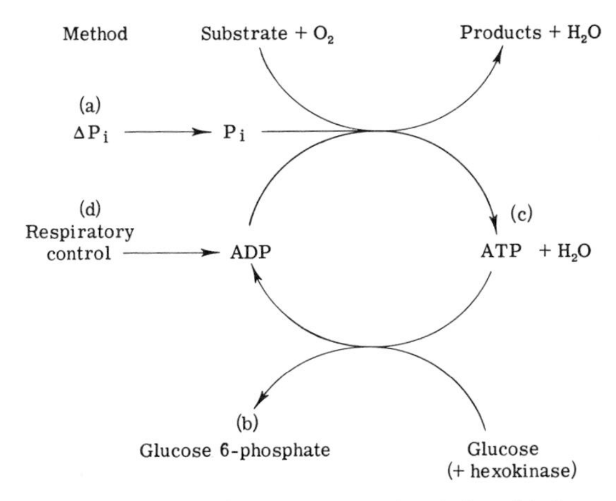

Fig. 1-2 Measurements of oxidative phosphorylation: (a) disappearance of P_i (colorimetric analysis); (b) formation of glucose 6-phosphate (radioisotope or enzymatic analysis); (c) formation of ATP (enzymatic analysis); (d) respiratory control (polarographic analysis).

respiratory control. However it is not applicable to the assay of individual phosphorylation sites and cannot be used when an ATPase is operative. Other procedures, either based on the assay of ATP by the firefly procedure or on enzymatic assays of glucose 6-phosphate formation, have several drawbacks, e.g., blank values due to the presence of adenylate kinase.

I should like to discuss now the experimental approach to oxidative phosphorylation. History of biochemistry has taught us that a clear understanding of complex multienzyme systems can be achieved only after the individual components have been separated and analyzed. At the time we started the resolution of oxidative phosphorylation, there were several reports describing the disruption of mitochondria into submitochondrial particles that catalyzed oxidative phosphorylation (Cooper and Lehninger, 1956; Ziegler *et al.,* 1956; Kielley and Bronk, 1958; McMurray *et al.,* 1958). In no case was there any indication for the resolution of soluble components.

First Approaches to the Resolution of the Membrane

My experience with the resolution of soluble multienzyme systems, such as the oxidative and reductive pentose phosphate cycle, has taught me that three ingredients are required for the resolution of a multienzyme system: (1) large quantities of stable starting material; (2) cheap labor; and (3) new ideas. When we started with this project about 15 years ago, large quantities of stable mitochondria could be prepared from bovine heart (Green *et al.,* 1956). Cheap labor was available in the form of a graduate student, Harvey Penefsky, who was willing to embark on this adventure. But we really had no new ideas, so we did what I call "instrumental research": You have no ideas, use a new instrument. At about that time, Dr. Nossal in Australia had designed a mechanical shaker for breaking up yeast cells with glass beads. So we used this machine to break up mitochondria. The first experiments were highly successful. After disruption of the mitochondria and differential centrifugation submitochondrial particles were isolated which oxidized substrates such as succinate, but did not generate ATP unless a soluble protein fraction (F_1) was added to the particles (Penefsky *et al.*, 1960). But soon thereafter experiments became difficult to reproduce. When Harvey Penefsky reported to me again and again that the data were "in the right direction" I knew we were in trouble. It was then that I recognized the need for a fourth ingredient when you work with membranes, namely a philosophy. I began to develop the "Tagfy" philosophy: "Troubles are good for you" which has sustained us ever since. I shall attempt to document its value during the rest of this lecture. Before doing so I must give credit to Dr. O. T. Avery who isolated a genetically active DNA from pneumococci and who foreshadowed the Tagfy philosophy when he said:

> *Lesson 4:* It doesn't matter if you fall down as long as you pick up something from the floor while you get up.

Since we had difficulties with the assay of oxidative phosphorylation we searched for an alternative approach to the problem. Over 30 years ago Lardy and Elvehjem (1945) had proposed with remarkable vision that there was a link between oxidative phosphorylation and the ATPase activity of mitochondria. In collaboration with Dr. M. Pullman we embarked on the isolation of this ATPase and observed that crude preparations of the soluble coupling factor (F_1) contained ATPase activity (Pullman *et al.*, 1958). Since we had troubles with the coupling factor assay and it was easier to measure the hydrolysis of ATP than oxidative phosphorylation, we decided to take a chance on the hypothesis that there is a link between phosphorylation and the ATPase and proceeded to purify the ATPase from the extract. However, the enzyme appeared to be extraordinarily labile and there were endless troubles in purifying it, until we discovered that the enzyme was labile at 0° but quite stable at room temperature. Not only was this the end of our troubles with purification, but the cold-lability served now as a tool to determine the relationship between the ATPase and the coupling factor. Is the coupling factor also cold-labile? The answer was yes, and the property which had delayed progress for months became a blessing. You can see, troubles can be good for you.

But a new complication arose when we made a systematic comparison of the purification of the ATPase activity and the coupling activity. The ratio of the two activities did not stay constant, but increased in favor of the ATPase activity. Was it possible that we were dealing with two different proteins, which were both cold-labile? We started to worry again until we realized that our recovery of the total ATPase activity was quite remarkable. In fact the final highly purified preparation contained twice as much ATPase than we had estimated to be present in the crude extract. It became apparent that there must be some factor present in the crude extract which inhibited ATPase activity. Thus a new research project was started and a protein inhibitor of the mitochondrial ATPase was found and purified (Pullman and Monroy, 1963). We shall return to this inhibitor later.

In Table 1-3 I have listed some of the properties of F_1. I shall discuss only one which caused us a lot of trouble. At the International Congress of Biochemistry in Moscow, I reported our findings on the soluble mitochondrial ATPase (Racker *et al.*, 1963). Dr. E. C. Slater, the chairman of the session, asked a very pertinent question: "Is the soluble ATPase sensitive to oligomycin?" When I answered that the purified enzyme was insensitive, Dr. Slater responded with: "In this case, this enzyme has nothing to do with oxidative phosphorylation." And Dr. Slater used to be my friend. Actually, he still is, I hope. I was aware of this serious discrepancy and set out to resolve it.

TABLE 1-3 **Properties of Coupling Factor 1**

MW: 360,000
Catalysis: ATP hydrolysis and formation (on membrane)
Subunit structure: 5 different subunits
Cold lability: Dissociation into subunits
Inhibitors: ATPase activity inhibited by sodium azide,
  but not by oligomycin or DCCD
Electron microscopy: 85 A spheres

Allotopic Properties of F_1

Our working hypothesis was simply this: F_1 is a mitochondrial protein and its properties might have been altered when it was separated from the membrane. It was therefore necessary to prepare a mitochondrial membrane free of F_1, and to find out whether oligomycin sensitivity can be conferred by reconstitution with pure F_1. I encountered considerable trouble when I attempted to remove all F_1 molecules from submitochondrial particles. I was encouraged by the observation that the cold lability of the soluble protein was not a property of the membrane-bound enzyme again suggesting a difference in properties. On the other hand, the stabilization of the enzyme by the membrane precluded a simple inactivation procedure. Following a suggestion by the physicist Leo Szilard, I treated submitochondrial particles with urea in the cold and this seemed successful at first. However, results were highly nonreproducible until I discovered that treatment of submitochondrial particles with trypsin markedly increased the ATPase activity. It soon became apparent that the protein inhibitor of the ATPase (Pullman and Monroy, 1963) was

TABLE 1-4 **Conferral of Oligomycin Sensitivity to F_1 by Addition to Resolved Particles**[a]

	ATPase activity (μmoles P_i formed in assay)		
	Without oligomycin	With oligomycin	Inhibition (%)
F_1	1.6	1.6	0
TU-particles	0.2	0.04	80
+ F_1	1.5	0.1	93
+ trypsin + F_1	1.5	1.4	8
+ F_1 + trypsin	1.6	0.5	70

[a]Experimental conditions were as described by Racker (1963).

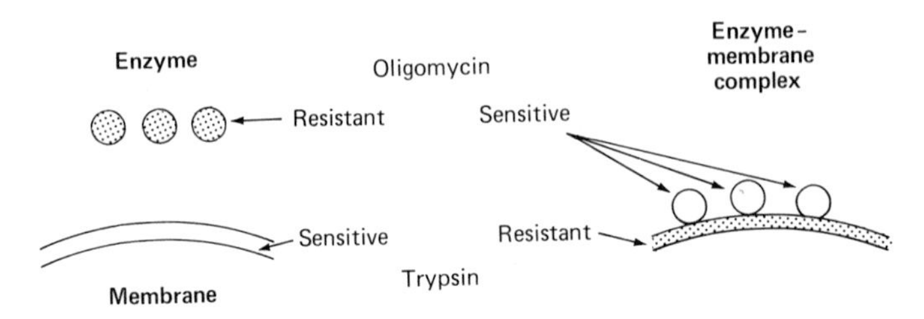

Fig. 1-3 The concept of allotopy.

very sensitive to proteolytic enzymes and that F_1 which was associated with the inhibitor was resistent to urea. After treatment with trypsin, F_1 became sensitive to urea and was removed from the membrane. Now we could perform the critical experiment shown in Table 1-4. The results were unambiguous. The ATPase activity of F_1 which was insensitive to oligomycin became sensitive after attachment to the membrane (Racker, 1963). Treatment of the F_1-covered membrane with trypsin prior to urea did not impair its ability to confer oligomycin sensitivity. However, as can be seen from Table 1-4 the membrane devoid of F_1 was very sensitive to trypsin, but became again more resistant by reconstitution with F_1. We thus arrived at the concept of allotopy (Racker, 1967) shown in Fig. 1-3: The properties of solubilized membrane proteins are different from those of the proteins bound to the membrane. This is illustrated by the change of the ATPase with respect to oligomycin sensitivity and cold lability. The membrane itself is altered by F_1 in its sensitivity to trypsin.

These experiments not only solved the dilemma of the oligomycin sensitivity of F_1 but opened up a new avenue of experimental approach. We now had the aid of a simple biological assay, namely, measurement of the oligomycin sensitivity of the ATPase, in our attempts to resolve membrane components. After many fruitless attempts, I performed a successful experiment. I must confess, however, that the secret of the success of the experiment was my absentmindedness. In preparing submitochondrial particles for sonication I forgot to add the usual buffer, which I am sure every graduate student or postdoctoral fellow would have added. I exposed mitochondria to sonic oscillation in a sucrose solution which was free of salts and centrifuged for 1 or 2 hours at 105,000 *g*. The clear yellowish supernatant contained not only the system required for rendering F_1 sensitive to oligomycin, but catalyzed, after addition of coupling factors, oxidative phosphorylation though at low efficiency. Had I really succeeded in solubilizing an active system catalyzing oxidative phosphorylation? When I studied the system more closely, it soon became apparent that this was not the case. On addition of buffers and salts which were required for oxidative phosphorylation, the clear solution clouded up and on centrifugation the activity was readily sedimented (Racker, 1964). This brings us to another lesson, particularly important in the light of subsequent claims regarding solubilized systems which catalyze oxidative phosphorylation.

> *Lesson 5:* Not everything that shines is gold. Not everything that floats after high-speed centrifugation is soluble. We want to make sure that the proteins we study are lean and don't carry fat as life preservers, which prevent them from sinking. If we claim solubilization too hastily, all that is sinking is our scientific reputation.

Electron Microscopy

These interesting observations motivated us to call upon electron microscopy for guidance and a collaborative effort was initiated with Drs. D. Parsons and B. Chance (Racker *et al.*, 1965). Trypsin-treated submitochondrial particles looked very much like untreated particles exhibiting the characteristic inner membrane spheres first discovered by Fernández-Morán (1962). However, treatment with 2 *M* urea in the cold which removed the ATPase also removed the spheres. This was the first indication that the inner membrane spheres are F_1 molecules. A picture of F_1 revealed characteristic 85 Å spheres shown in Fig. 1-4. We felt that for final proof we should show that F_1 added to TU-particles

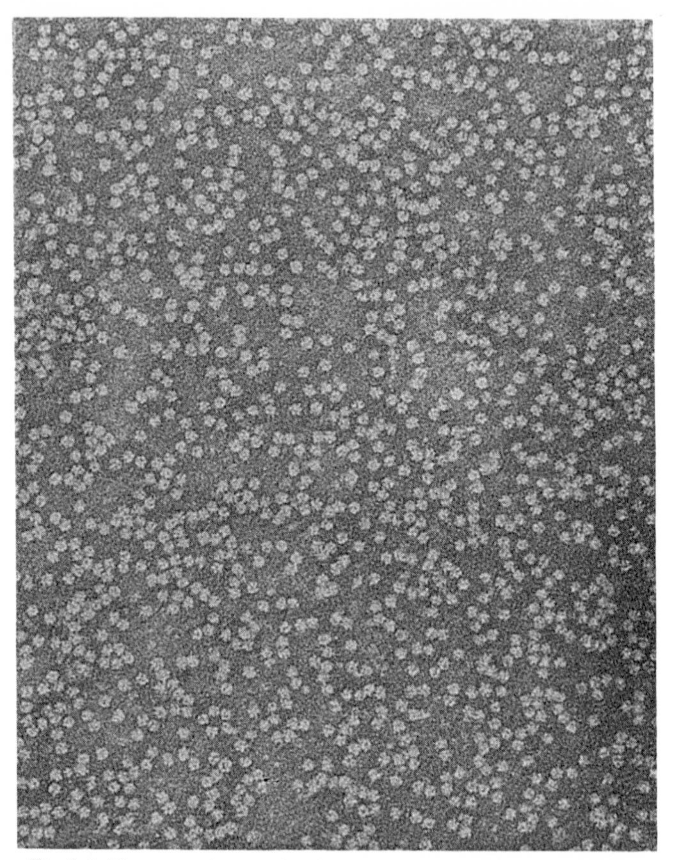

Fig. 1-4 Electron micrograph of mitochondrial ATPase (negative stain; magnification × 144,000).

yielded the typical appearance of active submitochondrial particles. But these experiments gave us trouble, particularly since the electron micrographs were variable and it was easy to fall into the pitfall of selecting fields. In desperation we resorted to statistics. I reconstituted various particles with and without F_1 and submitted numbered (unidentified) samples to Dr. Parsons who prepared the electron micrographs. He sent the pictures to Dr. Chance who employed a professional sphere-counter to evaluate the number of spheres per particles (Racker *et al.*, 1964). The result: It was in the right direction! We actually knew why we had so much trouble with the system but couldn't eliminate it. The trypsin-treated particles, although quite suitable for the binding of small quantities of ATPase and conferral of oligomycin sensitivity, were severely damaged and did not bind the massive amounts of F_1 required for oxidative phosphorylation or for obtaining convincing electron micrographs.

At about this time new troubles confronted us when Lee and Ernster (1966) reported that the P:O ratio of submitochondrial A-particles (Conover *et al.*, 1963) prepared by sonication of bovine heart mitochondria at an alkaline pH in the presence of EDTA, was markedly elevated by low concentrations of oligomycin (0.2 μg per mg particle protein). Since we had claimed that A-particles required F_1 and other coupling factors for oxidative phosphorylation, the authors were justified in suggesting that these so-called coupling factors may indirectly help to preserve the function of the membrane. Like oligomycin they may, for example, prevent the breakdown of a high-energy intermediate rather than serve as catalysts. We were not as upset by this development as we might have been because for some time we had known that A-particles were only partially depleted of F_1. Moreover, addition of an antibody against F_1 gave rise to an inhibition of phosphorylation which could no longer be restored by addition of oligomycin. It thus seemed clear that the coupling factor was contributing a function which oligomycin could not fulfill (Fessenden and Racker, 1966). When these data were first presented at a Gordon Conference, Dr. Ernster was impressed by the experiments with the antibodies but not entirely convinced. He assured me that we could really convince him of the special role of coupling factors if we could prepare submitochondrial particles completely free of F_1, which responded to F_1 but not to oligomycin.

Isolation of Resolved Particles

We again accepted the challenge and explored alternative methods to the treatment of particles with trypsin which we

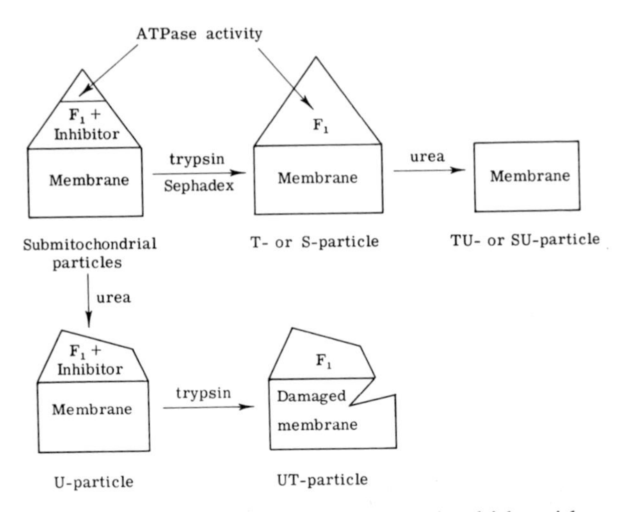

Fig. 1-5 Preparation of resolved submitochondrial particles.

knew damaged the membrane. Since the ATPase inhibitor which we had to remove before urea treatment is a small protein (11,000 MW), Mr. L. Horstman, my research assistant, explored the possibility of passing submitochondrial particles through a Sephadex column. He tried these experiments without any encouragement from me because I did not think that the procedure could be carried out without damage to the particles. However, it worked, which brings us to the next lesson.

> *Lesson 6:* Progress is made by young scientists who carry out experiments old scientists said wouldn't work.
>
> F. Westheimer

As shown in Fig. 1-5 the ATPase activity of the particles was greatly stimulated after passage through Sephadex G-50 and was rendered sensitive to urea as in the case of trypsin-treated particles. The figure illustrates also that treatment with trypsin gives rise to damaged submitochondrial particles. Reconstitution of Sephadex-urea treated particles with F_1 gave electron micro-

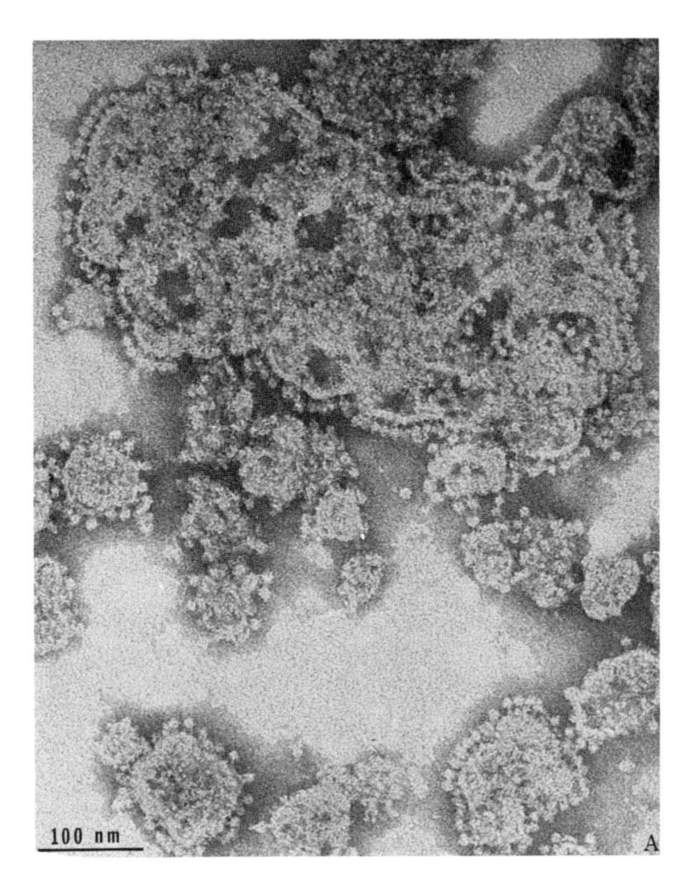

Fig. 1-6 (A) Electron micrograph of submitochondrial particles (negative stain; magnification × 144,000).

graphs that did not need statistical analysis. Figure 1-6(A) shows the characteristic appearance of submitochondrial particles lined with the 85 Å inner membrane spheres. In Fig. 1-6(B) are particles that have been passed through Sephadex and treated with urea. In Fig. 1-6(C) the same particles are shown after reconstitution with F_1.

Oxidative phosphorylation could be restored to the particles that have been treated with Sephadex-urea by addition of

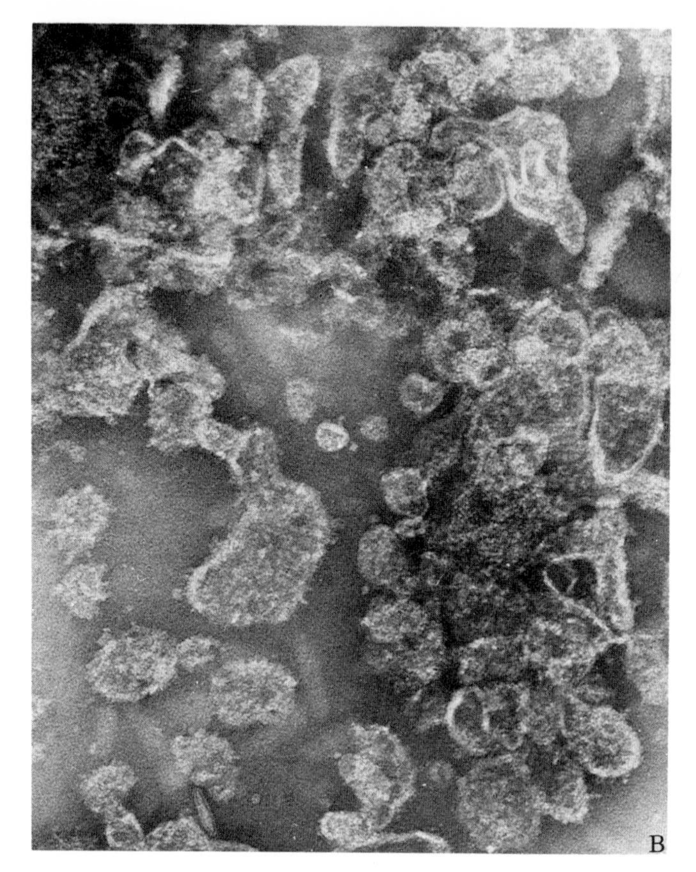

Fig. 1-6 (B) Electron micrograph of submitochondrial particles after treatment with Sephadex and urea (negative stain; magnification × 144,000).

coupling factors (Table 1-5). As requested by Dr. Ernster we could show that the coupling factors could not be replaced by addition of oligomycin (Racker and Horstman, 1967).

We thus decisively established the identity of F_1 with the inner membrane spheres and at the same time obtained evidence for a new concept. We realized that a membrane component such as F_1 has both a structural and catalytic role (Racker, 1967). The

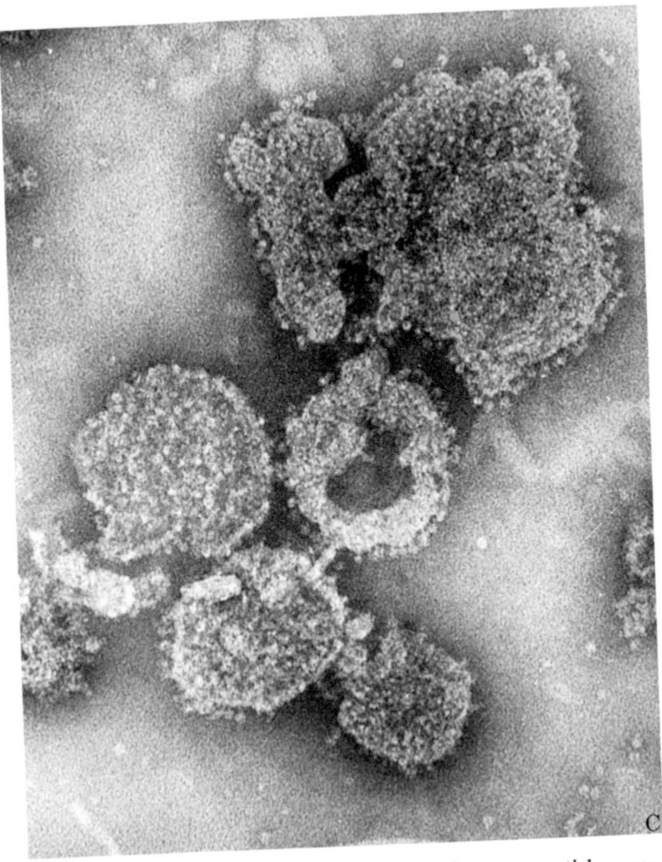

Fig. 1-6 (C) Electron micrograph of Sephadex-urea particles reconstituted with F_1 (negative stain; magnification × 144,000).

structural role could be replaced by low concentrations of either oligomycin (Lee and Ernster, 1966) or of dicyclohexylcarbodiimide (Racker and Horstman, 1967). Chemically modified F_1 which had no ATPase activity (Penefsky, 1967; Racker and Horstman, 1968) or yeast F_1 (Schatz *et al.*, 1967) could act as structural but not as a catalytic factor. Perhaps the most instructive experiment on this point was carried out by G. Schatz with

TABLE 1-5 **Reconstitution of Oxidative Phosphorylation to ASU-Particles**[a]

Additions	P:O
None	0.03
F_1, F_2, OSCP, F_6	2.8

[a]Experimental conditions were as described by Racker *et al.* (1969).

hybrid yeast particles and specific antibodies as shown in Table 1-6. Yeast submitochondrial particles which had a P:O ratio of about 0.5 with succinate were inhibited by an antibody against yeast F_1, but not against bovine F_1. Phosphorylation in all-beef particles (which Dr. Schatz referred to as kosher particles) was inhibited by anti-bovine F_1 but not by anti-yeast F_1. Bovine submitochondrial particles that contained insufficient amounts of bovine F_1 had low P:O ratios. Phosphorylation was stimulated by yeast F_1 but such particles were inhibited only by the anti-bovine F_1. We thus could differentiate between the catalytic activity of the bovine F_1 and the structural activity of the yeast F_1 by the insensitivity of the latter to antibody. I shall again discuss in the

TABLE 1-6 **Effect of Antibody on Reconstituted Submitochondrial Particles**[a]

	P:O		
	All beef	All yeast	Hybrid beef particles + yeast F_1
Control particles	0.53	0.39	0.58
+ anti beef F_1	0.06	0.35	0.12
+ anti yeast F_1	0.55	0.04	0.54

[a]Experimental conditions were as described by Schatz *et al.* (1967).

next lecture the structural and catalytic role of membrane components and a similar resistance of the structural function to antibody in the case of the spinach chloroplast coupling factor. It is not unreasonable that the addition of an antibody does not interfere with a structural role of a membrane component as long as the latter is not displaced from its position.

When we moved from New York City to Cornell University in Ithaca, we left behind an excellent electron microscopist. We couldn't find an electron microscope technician in Ithaca, a town of about 25,000 inhabitants, and I had no choice but to learn the trade myself. As is usual for a novice, I could see little in the microscope and my negatively stained preparations looked to me like dim shadows. When I was told that an English electron microscopist had used silicotungstate instead of phosphotungstate to obtain better contrast, I placed a rush order to England for a sample of silicotungstate. When it arrived (8 months later) I had already become an expert electron microscopist–I could see what I wanted to see. Nevertheless I tried the silicotungstate preparation and instead of getting better pictures I found that silicotungstate altered the membrane surface and removed the inner membrane spheres. Rather than discarding these disappointing observations, we decided to explore the biochemical basis of these morphological changes (Racker *et al.*, 1969). We found that silicotungstate inactivated rapidly the ATPase activity. Moreover, several other proteins were stripped from the membrane including succinate dehydrogenase. Further studies on the reconstitution of oxidative phosphorylation in silicotungstate-treated particles led to the discovery of a new heat-stable coupling factor F_6 (Fessenden-Raden, 1972a).

Reconstitution of Oxidative Phosphorylation

We encountered the final and most serious trouble when we attempted to reconstitute oxidative phosphorylation. After several generations of postdoctoral fellows and years of unsuccessful attempts, we decided to look at the problem from a different point of view. We had operated on the assumption that the mechanism of oxidative phosphorylation is similar to that of substrate level oxidative phosphorylation as illustrated by the mechanism of action of glyceraldehyde-3-phosphate dehydrogenase. By analogy this chemical formulation included two or more high-energy intermediates but no specific function for the membrane. On the other hand, Mitchell (1961, 1966) had proposed a chemiosmotic hypothesis which not only included a specific function for the membrane but an asymmetric organization of its constituents. Briefly, Mitchell proposed that the function of the respiratory chain is to translocate protons thereby establishing a proton motive force which consists of a ΔpH and a membrane potential. The oligomycin-sensitive ATPase functions as a proton pump in reverse, utilizing the proton flux to make ATP. We know that other biological pumps that are driven by ATP, e.g., the Ca^{2+} pump, can operate in reverse and utilize a Ca^{2+} flux to generate ATP.

If Mitchell is right, we argued, we must make compartments with phospholipids and assemble the respiratory chain asymmetrically. We embarked therefore on a systematic analysis of the topography of the respiratory enzymes and coupling factors in the inner mitochondrial membrane (cf. Racker, 1970a) and also analyzed the role and topography of phospholipids (Burstein *et al.*, 1971a,b). These experiments, which we shall discuss later in greater detail, led us to the realization of the importance of the asymmetric assembly of the membrane. However, when we returned to our attempts to reconstitute oxidative phosphorylation as it occurs in submitochondrial particles, we again met with failure. One of the major difficulties was caused by the fact that when we reconstituted respiratory enzymes with phospholipids, the enzymes preferentially assumed the position they occupy in mitochondria, namely, with the side that reacts with cytochrome *c* on the outside and the side that reacts with oxygen on the inside. You will hear in a later lecture how this orientation caused us a great deal of trouble and how we eventually overcame the difficulty. Consistent with the other happy endings I told you about today, these difficulties were turned into assets and even-

TABLE 1-7 **Troubles Are Good for You**

Troubles	Dividends
1. Difficult resolution	Easy reconstitution
2. Difficult F_1 assay	Discovery of ATPase
3. Cold lability of F_1	(a) No cold-room chills (b) Identification of coupling factor with ATPase
4. Variable ATPase/coupling factor ratio	Discovery of ATPase inhibitor
5. Resistance of soluble ATPase to oligomycin	(a) Assay for membrane components (b) Concept of allotopy
6. Stimulation of P:O by oligomycin	(a) Complete resolution of F_1 from particles (b) Concept of structural and functional role of membrane components
7. Lack of electron microscopist	Silicotungstate particles and discovery of F_6
8. Negative experiments on reconstitution of oxidative phosphorylation	Insight into topography of membrane and eventual reconstitution
9. Difficulties in isolation of pure components of the proton pump	Reconstitution of other pumps and new insight into mechanisms
10. Friends who challenge you	Friends who repeat your experiments

tually allowed us to study a variety of properties such as respiratory control and ion translocation in reconstituted systems. Finally, we reconstituted the ATP-driven proton pump of mitochondria (Kagawa and Racker, 1971; Kagawa *et al.*, 1973a) and then oxidative phosphorylation (Racker and Kandrach, 1973; Ragan and Racker, 1973a).

In the last Table (1-7) I have assembled the experimental evidence in favor of the Tagfy philosophy.

One of the most difficult problems working with membranes is the close association of its components with phospholipid and with other proteins. This makes separation difficult and requires a cautious approach to resolution and fractionation. On the other hand, once we can devise a mild procedure, which preserves the reconstitutive properties of the component, reassembly is simple and rapid.

The difficulties encountered with the assay of F_1 as a coupling factor were later traced to variabilities in the resolution and a role of cations for F_1 binding. By approaching the problem from a different direction, an independent assay for F_1 was found in the ATPase activity which allowed us to develop a more dependable and rapid assay procedure. The cold lability of F_1 had not only the dividends listed in the Table but represents an interesting physical-chemical property which we shall discuss in the next lecture.

Activity ratios have been often used in biochemical investigations to decide whether a single protein catalyzes more than one reaction. The possible pitfalls of such a procedure is illustrated by the finding that the ratio may vary because one of the activities can be influenced by external conditions without affecting the other, even though they are catalyzed by the same protein. It is quite likely that this will be true also for other multiheaded enzymes.

Since the allotopic properties of F_1 have been described, numerous additional examples for allotopy have been observed both in our and other laboratories. For experimental purposes perhaps the most important allotopic property is stability, which almost invariably diminishes when the protein is detached from the membrane. In some cases addition of phospholipids helps stabilization, a procedure which at times is valuable for preservation, but of little help for purification. Glycerol is often quite effective in stabilizing membrane proteins.

The startling discovery of Lee and Ernster (1966) that oligomycin stimulates phosphorylation in submitochondrial particles at first challenged the significance of our observations with coupling factors. But eventually it led to a new concept of the duplicity of roles played by various membrane protein components and to a better understanding of the mode of action of oligomcyin. Since it is now apparent that some of the proteins fulfill a structural as well as a catalytic function in the membrane, the question arises whether there is a need for a "structural protein" or whether all structural requirements can be met by enzymes and phospholipids.

Silicotungstate proved to be a very convenient reagent for the removal of surface protein from submitochondrial particles. As will be shown in the next lecture it acts similarly on chloroplasts.

Although as mentioned earlier we have reconstituted the proton pump of mitochondria, at the time some of the components were still impure. The purification of the hydrophobic membrane proteins gave us a lot of trouble. We therefore started several years ago to work on other pumps such as the Ca^{2+} pump of sarcoplasmic reticulum and the Na^+-K^+ pump of the plasma membrane. We have successfully reconstituted these pumps with highly purified preparations of the corresponding ATPases (e.g., Racker, 1972a). Work with these proteins has given us new insight in the mode of action of ion pumps.

Finally, let me remind you how often our work has been challenged by our biochemical colleagues. These challenges were welcomed. Many of my friends who have lectured in my presence have been exposed to similar challenges. I have obtained great satisfaction from convincing my friends who first were most skeptical and who are now successfully and often more competently repeating our experiments.

Lecture 2

Photophosphorylation

All flesh is grass.

Prophet Isaiah

Listening to both sides of a story will convince you that there is more to a story than both sides.

Frank Tyger

On the Origin of Life

You are all aware that the energy required for life is ultimately derived from the light of the sun with the aid of photosynthesis which takes place in chlorophyll-containing membranes. A molecular biological version of the origin of life runs something like this: On the first day God said, "let there be light," and there was light. On the second day God said, "let there be water," and there was water. On the third day God said, "Let there be membranes which contain chlorophyll and which can utilize the energy of light." On the fourth day the chlorophyll-containing vesicles made a ~ and the squiggle generated ATP, GTP, UTP, and CTP. On the fifth day God polymerized ATP, GTP, UTP, and CTP and created ribonucleic acid and God saw that it was good. On the sixth day the Lord said it was not good for RNA to be alone and he caused a deep sleep to fall upon RNA (he incubated at 0°) and he took a rib out of (rib)onucleic acid and made DNA from it (the rib was later renamed oxygen). And RNA and DNA were both naked but they were not ashamed. On the seventh day

God rested and that's when all the trouble started. There was the ~ with its bad influence and persuaded DNA to do what it was not supposed to do. DNA and RNA made hybrids and afterwards they felt ashamed because they were naked. So they covered themselves by making a coat protein and this is how the first virus was created.

God was very angry when he saw it and blamed RNA, and RNA blamed DNA and DNA blamed the ~ (this was the first case of passing the buck). But God did not like to have viruses float around in Eden and he banned RNA and DNA onto Earth where they lived unhappily ever after.

You may wonder why all this is not better known. Actually a paper on it was submitted to *The Journal of Biological Chemistry* but was rejected on two grounds: (1) Nobody can repeat the experiment; (2) Francis Crick had thought of it before.

Today I would like to talk about the biochemical approach to the problem of photophosphorylation. For many years, the study of photosynthesis was the domain of physicists. Light belongs to physics and biochemists wouldn't understand it. The physicists believed that the generation of bioenergy by light was very different from other biological systems of energy production.

Electron Pathway in Photophosphorylation

However, during the past ten years it has become very clear that the generation of ATP in chloroplasts is indeed very similar to oxidative phosphorylation. Although some differences exist which I shall mention later, the overwhelming impression which I wish to leave with you is the similarity of the two processes. As shown in Fig. 2-1 the flow of electrons through an oxidation chain generates in both systems the energy that is transferred to ADP. In both processes water serves as a hydrogen donor, but in mitochondria hydrogens are also derived from substrates such as pyruvate and fatty acids. Whereas in chloroplast the cleavage of water takes place with light and the aid of chlorophyll, mitochondria perform the task via the Krebs cycle. The electron

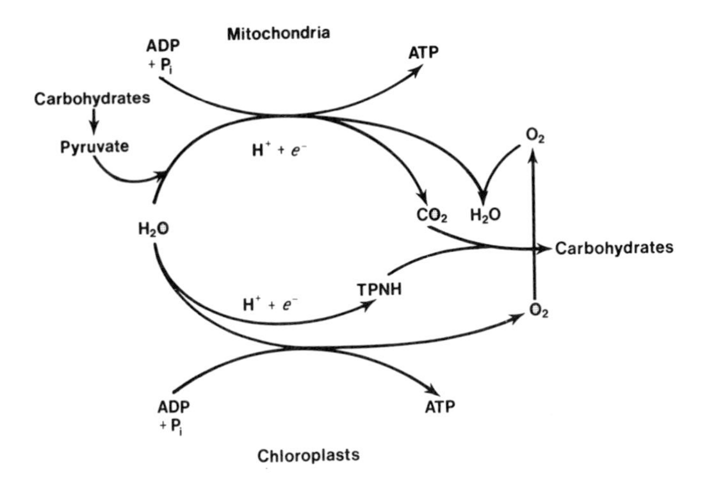

Fig. 2-1 Proton and electron flow in chloroplasts and mitochondria.

acceptor is different; it is oxygen in mitochondria and CO_2 (via TPN and the reductive pentose phosphate cycle) in chloroplasts. In Fig. 2-2 we focus closer on the oxidation chain in chloroplasts and at first sight we can see similarities with the respiratory chain in mitochondria. There are flavoproteins, nonheme iron proteins, quinones, cytochromes, copper proteins, and nucleotides in both systems to allow for the stepwise transport of electrons down the thermodynamic ladder. There are two "sites" of photophosphorylation which will be discussed later. A third site is missing;

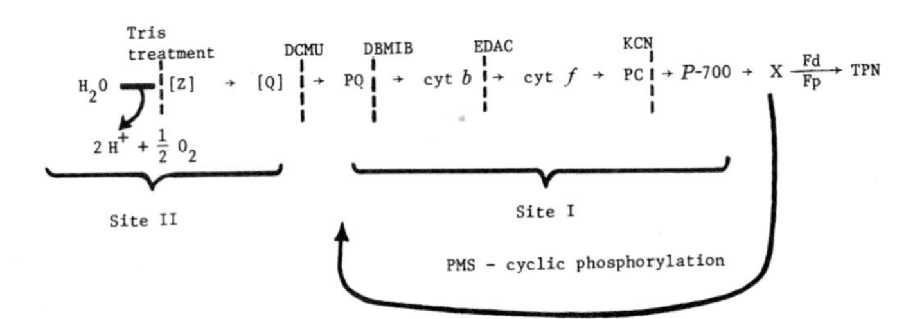

Fig. 2-2 The oxidation chain and phosphorylation in chloroplasts. The reaction sequence as outlined has been partly deduced from the effect of inhibitors that are listed on top of the individual steps (see abbreviations).

instead, a second light step is inserted at *P*-700 to raise the oxidation–reduction potential of electron flow to a level capable of reducing TPN. The energy is utilized for reductive purposes rather than for ATP formation.

The scheme shown in Fig. 2-2 is very tentative and there is no general agreement on the identity of the individual electron transport carriers. As in the case of mitochondria, inhibitors and physical resolution have aided in the development of the broad outline given in the figure. How metabolic inhibitors can be used in such systems and their limitations will be discussed in a future lecture. The position of the phosphorylation "sites" is also controversial, but there is almost general agreement that there are two "sites." According to the chemiosmotic hypothesis there is no sense in speaking of specific coupling sites of phosphorylation. Instead, we should recognize the segment responsible for the translocation of protons and identify the hydrogen carriers. The latter has not been achieved in either the mitochondrial or chloroplast systems; however, in both cases quinones are likely candidates. At the present time the simplest way of looking at photophosphorylation is that photosystem II catalyzes the cleavage of water and release of protons inside the chloroplasts. The proton motive force thus created is utilized by a proton pump that transports the protons from the inside to the outside with the simultaneous formation of ATP from ADP and P_i. Meanwhile, the electrons generated during water cleavage are shuttled to plastoquinone, which accepts protons from the outside and serves as a hydrogen carrier, delivering protons on the inside. Then the electrons move back in a loop to the outside reaching eventually, together with protons, TPN, forming TPNH, the hydrogen donor of the carbon cycle.

Cyclic phosphorylation, which we shall talk about in this lecture, takes place in the presence of an artificial electron carrier like phenazine methosulfate and appears to take place via Site 1 as indicated in Fig. 2-2. Since there is little general agreement on any of the above mentioned formulations, I have refrained from mentioning specific views or names but instead refer you to some review articles (Avron and Neumann, 1968; Bendall and Hill,

1968; Walker and Crofts, 1970; Siedow *et al.,* 1973; Trebst, 1974).

Further Similarities with Mitochondria

In Table 2-1 we can see some of the other similarities between the membranes of chloroplasts and mitochondria which are particularly apparent when we compare submitochondrial and subchloroplast particles. During electron transport, protons move from the inside to the outside in mitochondria. In chloroplasts, subchloroplast particles, and submitochondrial particles, protons move from the outside to the inside. This vectorial movement of protons requires an asymmetric organization of the oxidation chain which I shall discuss later. The chloroplast coupling factor CF_1 is very similar to F_1 in molecular weight, amino acid com-

TABLE 2-1 **Similarities between Oxidative Phosphorylation in Submitochondrial Particles and Photophosphorylation in Chloroplasts**

Properties	In submitochondrial particles and chloroplasts
Electron transport components	Cytochromes, flavoproteins, quinones, nucleotides
Proton movements during oxidation	From outside to inside
Coupling factor 1	Molecular weight, amino acid composition, subunit structure, cold lability are similar
ATPase activity	Masked but can be unmasked by trypsin with loss of coupling activity
Uncoupler, energy transfer inhibitors	FCCP, S-13 DCCD, Dio-9, quercetin
Exchanges	P_i–ATP, H_2O–ATP
Topography	Asymmetric organization
EM structure	Similar inner membrane spheres of 85–90 Å

position, and cold lability (cf. Racker, 1970b). The subunit structure, the Mg^{2+}-ATPase activity (Nelson *et al.*, 1972a), and the control of ATPase activity by an inhibitory regulatory subunit (Nelson *et al.*, 1972b) are additional features which they have in common. These similarities are particularly remarkable in view of the tremendous span of time that must have elapsed during the evolution from the spinach to the cow.

Chloroplasts, like mitochondria, exhibit the phenomenon of oxidation control, namely, the retardation of electron flux in the absence of ADP and P_i. Some uncouplers (e.g., FCCP) and energy transfer inhibitors (e.g., DCCD) affect both systems, though others (e.g., oligomycin) are not effective and some (e.g., Dio-9) are more effective in chloroplasts. Exchange reactions such as $^{32}P_i$–ATP, H_2O–ATP, etc., can be elicited in both systems though some operational differences can be observed. The morphology of submitochondrial particles and subchloroplast particles negatively stained with phosphotungstate is so similar that it is sometimes difficult to tell them apart with the color-blind eyes of the electron microscope. Like submitochondrial particles, the subchloroplast particles are lined with the characteristic 90 Å inner membrane spheres (Vambutas and Racker, 1965; cf. discussion in McCarty and Racker, 1966; Howell and Moudrianakis, 1967; Moudrianakis, 1968).

Some of these striking similarities have only become apparent in recent years. Earlier, the physicists could point to differences between oxidative phosphorylation and photophosphorylation such as the lack of an ATPase activity and of several exchange reactions. It was not until Petrack and Lipmann (1961) reported a light-dependent ATPase activity in chloroplast activated by dithiol compounds that attitudes started to change.

It was this discovery which aroused my interest in photophosphorylation. There were several other reasons why we embarked on a study of phosphorylation in spinach chloroplasts. (1) It was at a period in our research when we had a lot of trouble with oxidative phosphorylation and felt that an alternative approach to ATP generation may be fruitful. We hoped that the plant system was much simpler and we could avoid some of the complexities of the mitochondrial structure and of the three phosphorylation sites. (2) I had been working on the reductive pentose phosphate cycle of photosynthesis in spinach and was puzzled by some of the bioenergetic aspects of this process which did not quite add up. (3) We had learned to handle bushels of spinach which smelled better than bovine heart.

Resolution of a Coupling Factor

Our approach to the chloroplast problem was similar to that we had taken with mitochondria. A graduate student, Vida Vambutas, exposed chloroplasts to various treatments which had been successful in resolving coupling factors from mitochondria. After many failures, one procedure worked: Exposure of chloroplasts to sonic oscillation in the presence of 3% phospholipids yielded subchloroplast particles that catalyzed a very low rate of cyclic phosphorylation in the presence of phenazine methosulfate. When a soluble chloroplast extract was added to these particles, phosphorylation was markedly stimulated. In contrast to chloroplasts, which at that time could not be frozen without loss of activity, the subchloroplast particles obtained by this procedure were very stable and could be stored at $-70°$ for many months without loss of activity. The soluble coupling factor which we called CF_1 was stabilized by addition of ATP (Vambutas and Racker, 1965). Since this was all very reminiscent of the properties of F_1, we analyzed the coupling factor for ATPase activity but did not find any. At about that time I had found that the ATPase activity of submitochondrial particles was stimulated by treatment with trypsin (Racker, 1963). We therefore explored the effect of trypsin on CF_1 and observed an activation of a very rapid Ca^{2+}-dependent hydrolysis of ATP. About 600 μmoles of ATP were cleaved per hour per milligram chlorophyll which is of the same order as the rate of photophosphorylation. This rapid ATPase assay simplified the purification of CF_1. The purified protein, which served also as a coupling factor, could be stored like F_1 as a precipitate in 2 *M* ammonium sulfate at 0° for months without loss of activity.

More recently, the effect of trypsin on CF_1 has been studied in several laboratories. A brief exposure of illuminated chloroplast particles to trypsin (free of chymotrypsin) activated ATPase activity without interfering with coupling activity (Lynn and Straub, 1969). More extensive exposure to trypsin (Bennun and Racker, 1969) resulted in loss of CF_1 binding to the membrane as well as in the removal of the three small subunits (Deters *et al.*, 1975) that we shall discuss later.

Although CF_1 was stable as a precipitate in 2 *M* ammonium sulfate, it was, like F_1, very cold labile in solution in the presence of salts. We were interested in this curious property not only because it is a fascinating physical-chemical problem but because we hoped it may give us useful information about the mode of action of CF_1. We observed (S. Lien and Racker, 1971) that cold-inactivated CF_1 can be reactivated by incubating the enzyme at 30° in the presence of glycerol and ATP. After several hours up to 60% of the original activity was recovered as shown in Fig. 2-3. We are using this experiment as a starting point in the resolution and reconstitution of this enzyme from its subunits, hoping to get further insight into the specific role of the subunits.

The history of the coupling factor in chloroplasts goes back to experiments of Jagendorf and Smith (1962) who observed that chloroplasts, after washing with very dilute solutions of EDTA, catalyzed electron flow without phosphorylation. Avron (1963) restored phosphorylation to the washed chloroplasts by addition of the EDTA extract. Since he found that the stimulating factor in this extract was unstable, he made no attempts to purify or characterize it. A comparison of the factor in the EDTA extract with CF_1 revealed their identity (McCarty and Racker, 1966). Moreover, the simplicity of the extraction with EDTA lends itself as an initial step in the purification of CF_1 (C.-T. Lien and Racker, 1971).

The effect of EDTA at various concentrations on the loss of phosphorylating activity of chloroplasts and on the release of

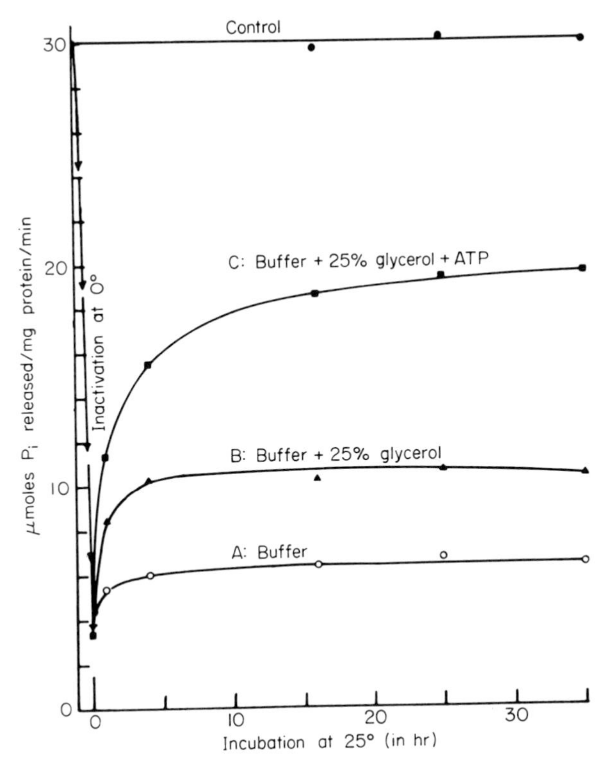

Fig. 2-3 Reactivation of cold-inactivated CF_1. Experimental conditions were as described by S. Lien and Racker (1971). Cold-inactivated CF_1 was incubated at 25° in the presence of the indicated compounds and assayed for ATPase activity.

ATPase is shown in Fig. 2-4. At low concentrations of EDTA (0.5 m*M*) CF_1 was released and photophosphorylation was lost, whereas at 5 or 10 m*M* EDTA the activity was preserved and little trypsin-activated ATPase was detected in the supernatant. Since with 0.5 m*M* EDTA and 10 m*M* NaCl the release of CF_1 was also prevented, it was apparent that ionic strength rather than high EDTA concentration was responsible for the protection. I mention these experiments because extraction of membrane components at low ionic strength is a very useful procedure which has

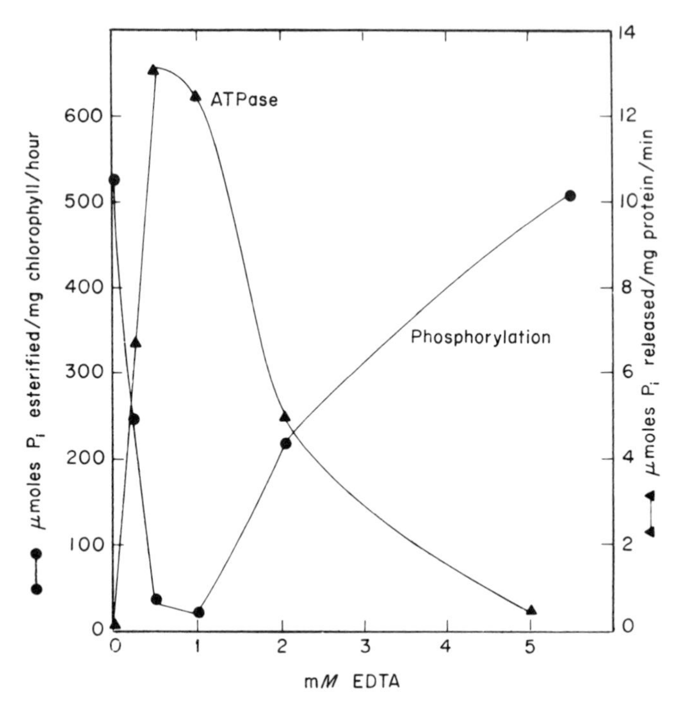

Fig. 2-4 Effect of EDTA on extraction of ATPase and phosphorylation activity. Experimental conditions were as described by McCarty and Racker (1966).

been applied to other systems including mitochondria and bacteria.

Oligomycin, which is a very effective energy-transfer inhibitor in mitochondria, has no effect on chloroplasts. In fact we have used oligomycin to see how much contamination of mitochondrial fragments are present in chloroplast fragments (Vambutas and Racker, 1965). An effective energy-transfer inhibitor in chloroplasts is Dio-9, an antibiotic which we obtained from the Netherland Royal Fermentation Industry in Amsterdam. Although Dio-9 is a compound of unknown structure and known impurity, it is a useful tool and potent inhibitor of photophosphorylation and of the ATPase activity of CF_1 (McCarty *et al.*, 1965). Like oligomycin in mitochondria, Dio-9 inhibits electron flow only in a coupled system but not in the presence of an uncoupler of photophosphorylation such as ammonium chloride (McCarty and Racker, 1966).

DCCD is a chemical of known structure and is readily available. It inhibits phosphorylation linked to electron flow in mitochondria (Beechey *et al.*, 1966) as well as in chloroplasts (McCarty and Racker, 1967). At very low concentrations it stimulates phosphorylation in partially uncoupled submitochondrial particles as well as in chloroplast fragments. As in mitochondria, electron flow is inhibited in chloroplasts by DCCD only when they are tightly coupled, exhibiting the phenomenon of oxidation control. In the presence of an uncoupler such as ammonium chloride, DCCD does not inhibit electron flow. In contrast to Dio-9, DCCD has little or no effect on the ATPase activity of soluble CF_1. These two energy transfer inhibitors therefore do not act at the same site.

Thus far we have seen little advantage in working with photophosphorylation, and most of the experiments I have described were first performed with mitochondria. I turn now to aspects of bioenergetics which reflect the advantages of the chloroplast system. For example, our recent work on the subunit structure of CF_1 was greatly aided by the fact that in contrast to F_1 which is a very poor antigen, the chloroplast factor and its subunits are quite antigenic in rabbits. Because of the antigenicity of CF_1 we decided to concentrate our efforts on the subunit structure of CF_1 (which we shall discuss in Lecture 4) rather than that of F_1 which is under investigation in several other laboratories (Senior and Brooks, 1970; Catterall and Pedersen, 1971; Lambeth and Lardy, 1971; Knowles and Penefsky, 1972). CF_1 antiserum inhibited all functions involving either the hydrolysis or generation of ATP while a control serum obtained from the same rabbit prior to immunization had no effect. Cyclic and noncyclic phosphorylation as well as light-activated ATPase activities were inhibited, but electron flow was not (McCarty and Racker, 1966).

Proton Movements in Chloroplasts

Proton movements were first discovered in chloroplasts by Jagendorf and Neumann (1965) who observed light and electron transport dependent uptake of hydrogen ions, which were measured by monitoring the pH of an unbuffered chloroplast suspension. The uncoupler-sensitive proton movements were maximal under conditions that were optimal for the formation of a high-energy state that drives ATP formation in the dark. Since energy transfer inhibitors such as DCCD did not inhibit light-driven proton translocations, the reaction was not expected to require CF_1. In fact, an antibody against CF_1 had no effect on proton translocation (McCarty and Racker, 1966). Yet as shown in Fig. 2-5, CF_1 was required for this process. Chloroplast particles depleted of CF_1 did not show the characteristic pH rise on illumination and complete restoration of the phenomenon was achieved by addition of CF_1 and Mg^{2+}.

We were faced here with an apparent paradox. We believe that CF_1 catalyzes the final transphosphorylation step in the formation of ATP from ADP and P_i. It should not therefore be required for the light-driven proton translocation which requires neither ADP nor P_i. The resistance of the process to energy transfer inhibitors and to CF_1 antibody were experimental observations consistent with that notion. Why then is the presence of CF_1 still needed?

This was one of the first experiments which led us to the concept of the dual roles of membrane components which I mentioned in the first lecture: CF_1 has a catalytic and a structural function. The structural function is not impaired either in the presence of the antibody or after chemical modification which results in a loss of catalytic activity. In some instances, e.g., when sufficient catalytically active CF_1 is available, energy-transfer inhibitors such as DCCD can fulfill the requirement of the structural component, explaining the stimulation of phosphorylation by DCCD in partly depleted fragments, and its ineffectiveness in completely resolved systems. It may be significant that the amount of energy transfer inhibitor required for this

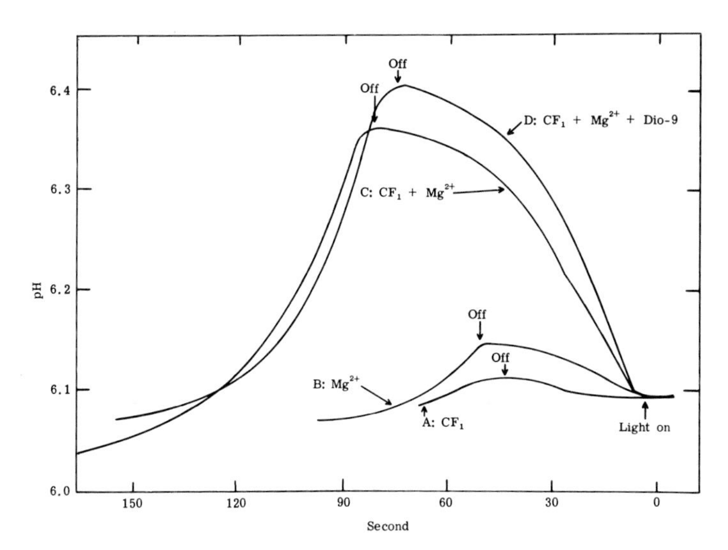

Fig. 2-5 Proton movements in resolved and reconstituted chloroplasts. Chloroplasts were resolved by EDTA treatment (A) with CF_1 added alone, (B) with Mg^{2+} added alone, (C) with CF_1 and Mg^{2+} added, and (D) both added as well as Dio-9. Experimental conditions were as described by McCarty and Racker (1968).

function occupies about the same space as normally held by the coupling factor, pointing to the possibility that certain regions on the membrane must be covered to permit function. These considerations are in line with the concept that oligomycin, DCCD, as well as F_1 (structural role) serve to render the membrane impermeable to protons.

The unambiguous elucidation of the participation of a coupling factor requires a complete resolution of the protein from the membrane. Just as in the case of submitochondrial particles, subchloroplast particles could be completely resolved with regard to CF_1 by treatment with silicotungstate (S. Lien and Racker, 1971). Silicotungstate-treated particles have no inner membrane spheres and do not phosphorylate even in the presence of low concentrations of DCCD. This is in contrast to the partially depleted EDTA-chloroplasts or the subchloroplast particles prepared in the presence of phospholipids. On addition of CF_1 to silicotungstate-treated particles, both morphological and functional restoration were achieved. However, the rates of the restored photophosphorylation were considerably lower than in reconstituted EDTA-particles. Moreover, as in the case of mitochondrial STA-particles, electron transport functions of the treated chloroplast particles were damaged also. We should therefore continue with attempts to prepare less damaged but fully resolved subchloroplast particles.

Another important first in bioenergetics was the demonstration of ATP formation in chloroplasts driven by imposing a ΔpH. Jagendorf and Uribe (1966) demonstrated that, during a transition from an acid to an alkaline pH, chloroplasts form ATP in the dark. We promptly confirmed this important finding and showed that the reaction was inhibited by anti-CF_1 or by Dio-9 (McCarty and Racker, 1966). It therefore became clear that this mode of ATP formation is related to photophosphorylation and involves the same components of the coupling device. I shall return to these experiments involving proton movements in the next lecture when we discuss the chemiosmotic hypothesis.

Asymmetric Assembly of the Chloroplast Membrane

One other and important similarity between chloroplasts and mitochondria is the asymmetric organization of electron transport catalysts within the membrane. We were first faced with this problem when we examined the role of plastocyanin in photophosphorylation. Plastocyanin has been proposed to serve as an electron carrier between photosystem I and II (Katoh and Takamiya, 1965). However, considerable controversy has arisen about its function, and in several papers a stimulation of electron transport by plastocyanin added externally to chloroplasts has been described. Yet externally added plastocyanin had no effect on phosphorylation. When we removed plastocyanin from chloroplasts by sonic oscillation we observed loss of phosphorylation. Addition of plastocyanin externally to these deficient subchloroplast particles stimulated electron flow to photosystem I (Hauska *et al.*, 1970), but again no phosphorylation was associated with this process. However, when sonication was performed in the presence of an excess plastocyanin, the resulting particles catalyzed photophosphorylation at a rate close to that of the untreated control (Hauska *et al.*, 1971; Racker *et al.*, 1972).

Experiments with antibodies against plastocyanin, cytochrome *f* and purified preparations of *P*-700 have led us to conclude that the electron transport components of the chloroplast membrane are organized asymmetrically just like those of the mitochondria as shown in Fig. 2-6. According to this scheme, 4 protons are generated on the inside, 4 protons consumed on the outside. Electrons move from system 2 to a quinone which accepts protons and while releasing these protons on the inside transfers electrons to the cytochrome–plastocyanin chain. With the aid of the second light reaction these electrons are conducted via the *P*-700-ferredoxin–flavoprotein system to TPN.

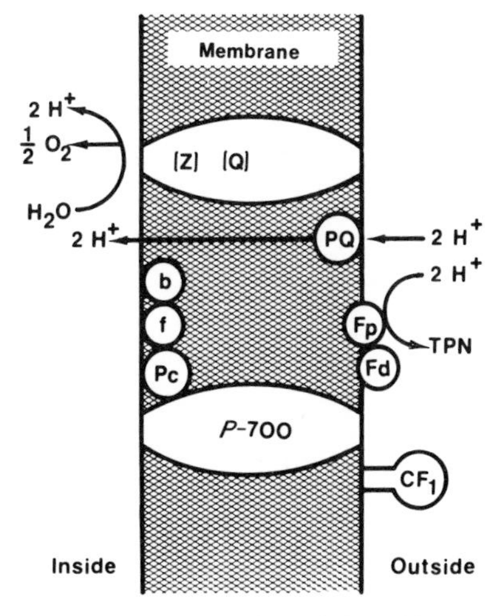

Fig. 2-6 The asymmetric organization of the oxidation chain in chloroplasts. In the primary photo event, which releases two protons, two unidentified components Z and Q (presumably a quinone) are believed to participate.

Having dealt until now with phenomena that can be observed with either chloroplasts or mitochondria, I should like to use the last part of the lecture to point to some interesting differences in the bioenergetics of these two organelles.

As mentioned earlier, unlike mitochondria which move protons out, chloroplasts take protons up. It is therefore not surprising that other ion movements and the response to ionophores are also different in these two organelles. However, these differences disappear when we compare subchloroplast and submitochondrial particles which both catalyze proton uptake. Thus the observed differences are not caused by a difference in fundamental mechanisms, but in the morphological organization of the two organelles.

Reversal of Photophosphorylation

Another difference, also not fundamental, but interesting from the physiological as well as theoretical point of view, is the lack of reversal of the energy transfer reactions in chloroplasts. The early difficulties encountered in demonstrating ATPase activity or $^{32}P_i$–ATP exchange in chloroplasts were in fact used by the physicists as an argument against the operation of the mitochondrial mechanism in chloroplasts. After all, according to the physicists all chemical reactions are reversible. I cannot resist to take this opportunity to digress for a moment and to give you my views on the problem of reversibility. Although the physicists may be theoretically correct, there are many enzyme-catalyzed reactions that to all practical purposes are irreversible. At least one cannot detect reversal with very sensitive methods and certainly one can exclude a physiological significance in the reverse direction.

Lesson 7: Some reactions are more reversible than others. If it takes a few years to demonstrate reversal it's not a practical experiment. I recommend that when an enzyme-catalyzed reaction tested under optimal conditions shows no sign of reversal by the time bacteria start growing, we should call the reaction irreversible.

It is possible, however, to induce ATPase activity (Petrack and Lipmann, 1961) and $^{32}P_i$–ATP exchange (Carmeli and Avron, 1966; McCarty and Racker, 1966) in chloroplasts. However, in both cases there is a requirement for a dithiol compound which is not needed in the forward reaction of photophosphorylation. We propose that the differences in the reversibility of energy transfer reactions in mitochondria and chloroplasts are related to the properties of the ϵ subunit of CF_1 (inhibitor of ATPase activity). Purified F_1 catalyzes a rapid ATP hydrolysis because the inhibitor subunit has been dissociated from the protein during fractionation. In contrast, the chloroplast coupling factor is inactive as ATPase, because it contains a firmly bound inhibitor (ϵ subunit), which is very hydrophobic (Nelson *et al.*, 1972b), but readily destroyed by treatment with trypsin. The following experiments favor the view that this masking of the ATPase is also responsible for the lack of reversibility in chloroplasts as measured, e.g., by the $^{32}P_i$–ATP exchange. Exposure of CF_1 to large amounts of dithiothreitol activates the ATPase activity. In contrast to trypsin-treated CF_1, which is no longer active as a coupling factor, dithiothreitol-treated CF_1 functions as a coupling factor when added to depleted chloroplasts. However, we have shown (McCarty and Racker, 1968) that the dithiothreitol-activated ATPase activity disappears when the protein is returned to the membrane. It is therefore apparent that on interaction with the membrane the protein undergoes a conformational change which induces a kinetic irreversibility of its enzymatic activity. An active ATPase with coupling factor activity can also be prepared in the presence of low concentrations of DTT if the chloroplasts are first exposed to light (McCarty and Racker, 1968). The presence of ATPase activity in CF_1 isolated from light-exposed chloroplasts also indicates a conformational change in the membrane-bound CF_1. More direct evidence for light-induced conformational changes in membrane-bound CF_1 was demonstrated by Ryrie and Jagendorf (1971) who observed uptake of tritium from water into CF_1 following exposure of chloroplasts to light.

The irreversibility of CF_1 action and the conformational plasticity of the protein may have interesting physiological implications. Whereas in mitochondria, reversal of electron flow participates in the reduction of DPN and TPN, in chloroplasts these reactions seem to be lacking.

In this lecture, I have attempted to show that in spite of the very basic difference in the primary event of photosynthesis and the cleavage of water by light in the presence of chlorophyll, the principles of ATP generation are the same in chloroplasts and in mitochondria. We have also seen that by using chloroplasts we have been able to use some approaches, e.g., based on the antigenicity of CF_1 and its subunits, which we were unable to apply to bovine F_1. An additional revenue of the approach of comparative biochemistry is further insight in the process of evolution. The remarkable similarity between the coupling factors from mitochondria, yeast, and spinach emphasizes that energy production linked to electron flow was discovered many million years ago and that nature (like scientists), on discovering something real good, repeats it over and over again.

Lecture 3

Functions and Structure of Membranes and the Mechanism of Phosphorylation Coupled to Electron Transport

> There is a complicated hypothesis which usually entails an element of mystery and several unnecessary assumptions. This is opposed by a more simple explanation which contains no unnecessary assumptions. The complicated one is always the popular one at first, but the simpler one, as a rule, eventually is found to be correct. This process frequently requires 10 to 20 years. The reason for this long time lag was explained by Max Planck. He remarked that "Scientists never change their minds, but eventually die."
>
> **J. H. Northrop**

> Man's mind stretched to a new idea never goes back to its original dimension.
>
> **Oliver Wendel Holmes**

General Comments

Oxidative phosphorylation and membranology are two areas of research which have collided rather recently. Although neither field has been moving very fast, the impact of the collision was very great. The reason for this can be traced to the remarkable differences in the personalities who were involved in that collision. A membranologist may be a physicist, certainly a physical chemist, sometimes an electrophysiologist, always an electron microscopist and a photographer, but above all he must be a man of decision so that he can select from 100 electron micrographs the one he likes. The investigator of oxidative phosphorylation often referred to as a mitochondriac, must be a versatile biochemist and must have a lot of patience to separate and purify complex mixtures of phospholipid and insoluble proteins. Above all he must be a man of imagination so he can see intermediates whether they are there or not.

It is indeed curious that for many years the biochemists working on oxidative phosphorylation completely ignored the fact that they were working with organized, vesicular structures. The physical chemists working on membrane proteins were, and I am afraid many still are, unconcerned whether these isolated preparations are still capable of physiological function. The electron microscopists were often engaged in studies of membranes with little or no knowledge of the biological significance of the structures they described. After the collision mentioned above, things were a little confusing, particularly when men of decision started to collaborate with imaginative biochemists. Some textbooks have still to recover from the fairy tales which appeared during that stormy period.

However, the last few years have brought about a refreshing change in atmosphere. The fog of polemics has cleared and although the sun is not exactly shining, the fruitfulness of the collision is visible and documented in the appearances of books, reviews, symposia, and articles on membranes that include aspects of both structure and function. Before discussing the mitochondria I should like to comment on some of the broader problems of membranes.

Function and Structure of Membranes

The Primary Function of Membranes

There can be little doubt that the primary and most important function of membranes is the formation of a compartment.

Membranes not only protect the cell from a hostile environment, but permit it to retain ions, coenzymes, enzymes, and other ingredients essential for life. Moreover, membranes separate intracellular compartments that fulfill specific functions such as biosynthesis, reproduction, and energy generation. This separation is essential for the autonomy and control of differentiated functions. It was a membrane and not Caesar who first said "divide and conquer."

The Secondary Function of Membranes

With the privilege of isolation comes the responsibility of communication. For this purpose the impermeable membranes acquired, during evolution, channels and carriers that allow for communication between the intracellular compartments as well as between the cell and its environment. As shown in Fig. 3-1 we can readily distinguish between the two extremes of the communication system: (a) the channel (or pore) which permits a very rapid means of transport, e.g., of ions, and is not greatly influenced by the state of the membrane fluidity, and (b) the mobile carrier that travels from one side of the membrane to the other and is therefore markedly influenced by the state of fluidity of the membrane. Somewhere in between these two extreme models is a variety of compromises such as the rotating carrousel

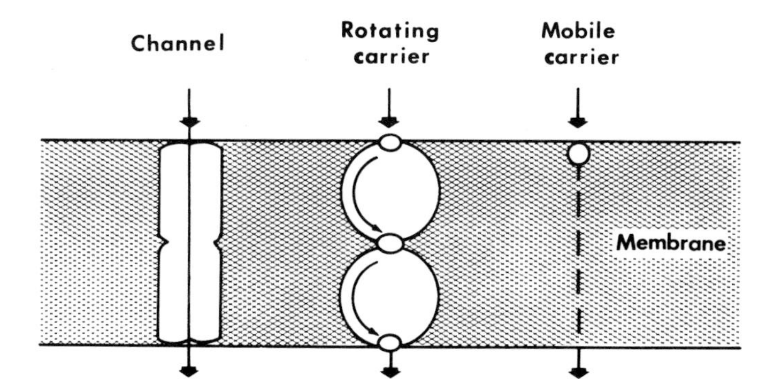

Fig. 3-1 Some communication systems in membranes.

model in the center of Fig. 3-1. We can mention gramicidin, alamethicin, and the proton pump of bacterial rhodopsin as well-established examples for the channel model, and valinomycin, nigericin, and many proton ionophores as prototypes for mobile carriers. Although there is no authentic example for the carrousel model, we cannot rule it out at the present time. However, as I shall discuss later, we did eliminate in a reconstituted system the operation of a particular carrousel model that has been proposed for the operation of the Ca^{2+} pump of sarcoplasmic reticulum.

The Tertiary Function of Membranes

In the course of evolution additional functions such as phagocytosis, pinocytosis, and accessory structures required for mobility were acquired by membranes which greatly widened the scope of membrane functions. Moreover, multienzyme systems, which can operate with greater efficiency because of proximity between the catalysts, have been assembled within the membranes. Important enzymes such as monoamine oxidase were anchored in the outer mitochondrial membrane at a safe distance from the receptor sites that respond to biogenic amines. Compartments were created that sequester cations such as Ca^{2+} which serves as secondary messenger controlling many metabolic events and which initiates muscular and nervous events.

What about the Structure of Membranes?

Most membranologists agree that membranes differ both in function and structure from each other and that each membrane is an individual. But the question is: Are there some common denominators? Are all membranes created equal? Is there something like a primary membrane structure common to all? Is there a basic similarity in the interplay between the proteins and lipids? Is there a unit membrane structure? Is there some common structural protein? The controversies on these points have simmered down and there is some consensus emerging that the common denominator of most, if not all membranes, is a phospholipid bilayer. This structure is well suited to fulfill the primary function of membranes–the formation of compartments. The original model of a pure phospholipid bilayer (Danielli and Davson, 1935) satisfied the need for compartmentation but required modification because of the need for communication. Singer and Nicolson (1972) have formulated a model consisting of a mosaic of phospholipids and proteins which accounts for the presence of transporters within the membrane. These authors have stressed the importance of fluidity of membranes which was documented by a variety of studies such as those of Fry and Edidin (1970), Kornberg and McConnell (1971), and Cone (1972).

The associations between membrane components vary greatly. Some protein components are very hydrophobic and are difficult to dissociate from the membrane; others are quite water soluble once they are dissociated from the other components. As a rule the hydrophobic components are located centrally, the hydrophilic component peripherally. Even within a single enzyme complex such as cytochrome oxidase which spans across the membrane, the less hydrophobic polypeptides are peripheral and the more hydrophobic subunits are internal (cf. Schatz and Mason, 1974). But even among the phospholipids we can differentiate

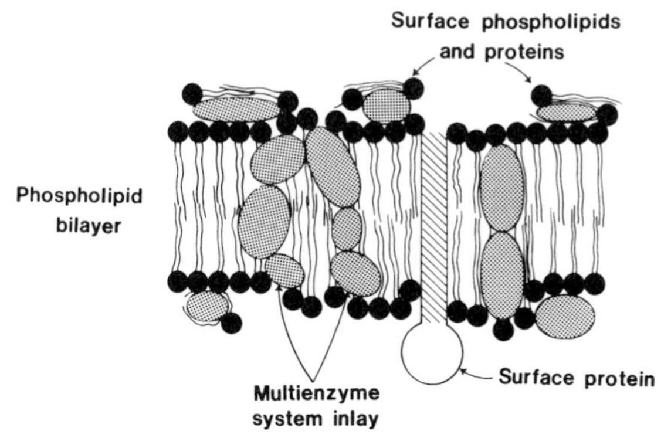

Fig. 3-2 Membrane structure.

between peripheral components that are accessible to the action of water-soluble phospholipase and internal lipids that are not. Surprisingly, in the case of the mitochondrial membrane, a relatively large fraction of the phospholipids can be digested without loss of the phosphorylating capacity (Burstein *et al.,* 1971a,b). These surface phospholipids appear to be important for ion translocation. We therefore suggest that the structure of membranes is still more complex than the currently accepted model and a tentative scheme is shown in Fig. 3-2, which includes surface phospholipids and proteins that can be digested or removed without affecting the basic membrane structure. These components (like F_1) are referred to as peripheral or extrinsic, whereas components that cannot be removed without disruption of the membrane structure are referred to as central or intrinsic (or integral).

There are two terms I have mentioned that require further discussion. What is a hydrophobic protein and what do we mean by the fluidity of a membrane? Unfortunately, we cannot define the hydrophobicity of a protein simply in terms of the number of its hydrophobic amino acid residues. This may be a contributing factor, but the tertiary structure is a much more important feature than the primary amino acid composition as demonstrated by the fact that we can convert a very hydrophilic protein into a hydrophobic protein by heat denaturation. More convincing is that a very hydrophobic protein, soluble in chloroform–methanol, can be converted into a highly water soluble protein and reconverted into a hydrophobic protein (Folch-Pi and Stoffyn, 1972). Another interesting example is CF_1, a hydrophilic peripheral membrane component composed of five very hydrophobic subunits (Nelson *et al.,* 1973). I wish we had a better insight into the forces that dominate protein structure and subunit assembly; it would not only increase our understanding of protein assembly but aid us in the purification of hydrophobic proteins. The hydrophobicity of membrane proteins and their great sensitivity to solvents has been a stumbling block that has greatly delayed progress in the isolation of pure internal membrane proteins.

The concept of fluidity has also suffered from semantic confusion. There is no generally accepted definition of this term and its use has been at least as fluid as the membrane itself. We can use precise physical terms such as transition temperature which can be applied to artificial membranes made of well-defined synthetic phospholipids. In natural membranes with mixtures of lipids and proteins, the responses to temperature are too complex to be useful in a definition. Enzyme activity of pure proteins that contain no phospholipids may show discontinuous responses to temperature and there are therefore no generally accepted methods to measure fluidity of natural membranes. Artificial probes are useful (Kornberg and McConnel, 1971; Shinitzky and Inbar, 1974) provided that they are truly neutral referees and do not preferentially interact with one or the other specific membrane component. Moreover, there is need for further sophistication regarding both the direction of motion and measurements that are influenced by the structure of the moving probe. For example, the horizontal or lateral mobility of a phospholipid (Kornberg and McConnell, 1971) or of a protein (Frye and Edidin, 1970) differs greatly from the movement of a polypeptide ionophore such as valinomycin which carries an ion from one side of the membrane to the other (Krasne *et al.*, 1971). Moreover, mammalian rhodopsin may exhibit considerable horizontal mobility (Cone, 1972) while bacterial rhodopsin is immobilized (Oesterhelt and Stoeckenius, 1971). The mobility of certain ionophores such as nigericin may serve for a quantitative evaluation of the "vertical fluidity" of a membrane as has been shown with the aid of a reconstituted proton pump (Racker and Hinkle, 1974; LaBelle and Racker, 1976).

Mechanism of Phosphorylation Coupled to Electron Transport

The catalysts of oxidative phosphorylation are seated in the inner mitochondrial membrane. There are three phosphorylation sites shown in Fig. 1-1. The first site is between DPNH dehydrogenase (Fp) and Q_{10}, the second between cytochrome b and c_1, the third between cytochrome c and oxygen. We show the

first site between the flavoprotein and Q_{10} although most textbooks have placed it between DPNH and the flavoprotein. You may wonder why and the answer is simple: Because Britton Chance said so. A few years back one of my graduate students, Peter Hinkle, made some observations which were incompatible with this view. My first attempt to convince Dr. Chance of the new location of site 1 phosphorylation by telephone was not successful. Peter Hinkle and I therefore traveled to Philadelphia and, after receiving various psychological and electrical shocks in the dark laboratory of Dr. Chance, further experimentations brought additional evidence in favor of the location of site 1 phosphorylation between the flavoprotein and Q_{10}. A paper in collaboration with Dr. Chance was published (Hinkle *et al.*, 1967b) and I hope (a) that we have really convinced him and (b) that the new assignment for site 1 will now find its way into the textbooks. After all, Britton Chance says so!

> *Lesson 8:* When you find something that disagrees with a statement in a textbook or in a paper or review, you have a choice of three courses of action. You can fight, or you can switch, but I would rather recommend the third possibility, that is, to get in touch with the person with whom you disagree. If the issue is indeed important to you, straighten it out by collaborative experiments rather than by polemics.

Figure 1-1 illustrates another important feature of oxidative phosphorylation. We can enter the oxidation chain at various points and thereby can examine a smaller segment of the pathway. With succinate in the presence of rotenone we eliminate operation of site 1 and measure the span between succinate and oxygen. With ascorbate-PMS in the presence of antimycin we measure only the span between cytochrome *c* and oxygen. With DPNH as hydrogen donor and Q_1 as hydrogen acceptor, the first site can be measured with other reactions eliminated by the presence of antimycin A. The first site can also be evaluated in the reverse direction by measuring the energy-dependent reduction of DPN by succinate.

I want to discuss now current views on the mechanism of oxidative and photophosphorylation. There are two fundamentally different hypotheses, the chemical and the chemiosmotic. There are also references to a conformational hypothesis. However, conformational changes are essential features of every process that involves enzyme catalysis and the term may therefore be confusing. We shall discuss the role of conformational changes in the context of the chemical as well as the chemiosmotic hypothesis.

The Chemical Hypothesis

It was proposed over twenty years ago (Slater, 1953) that oxidative phosphorylation proceeds analogously to the reaction catalyzed by glyceraldehyde-3-phosphate dehydrogenase via a high energy-intermediate A~X which contains no phosphate and is a derivative of a respiratory catalyst. The chemical hypothesis included also the conversion of A~X to X~Y, a high-energy intermediate that does not involve a member of the respiratory chain. A~X or X~Y are transformed by phosphorolysis to X~P, from which the phospate group can be transferred to ADP to form ATP. There are several good reasons (Racker, 1965) why X~Y, a high-energy state or intermediate, which is independent of the oxidation chain, was postulated and at one time was even included in the chemiosmotic hypothesis (Mitchell, 1966). Additional compelling evidence has emerged from studies with yeast and with reconstituted systems. Yeast promitochondria lack an oxidation chain but still perform work in the presence of ATP (Groot *et al.*, 1971). Similarly, phospholipid vesicles reconstituted with the oligomycin-sensitive ATPase, which are incapable of respiration, catalyze a $^{32}P_i$–ATP exchange which requires a high energy state or intermediate (Kagawa and Racker, 1971). Thus, X~Y can be produced from ATP in the absence of an active respiratory chain. On the other hand, the high-energy intermediate or state can be generated by oxidative energy and utilized for several energy-dependent reactions under conditions that rule out participation of ATP (Fig.

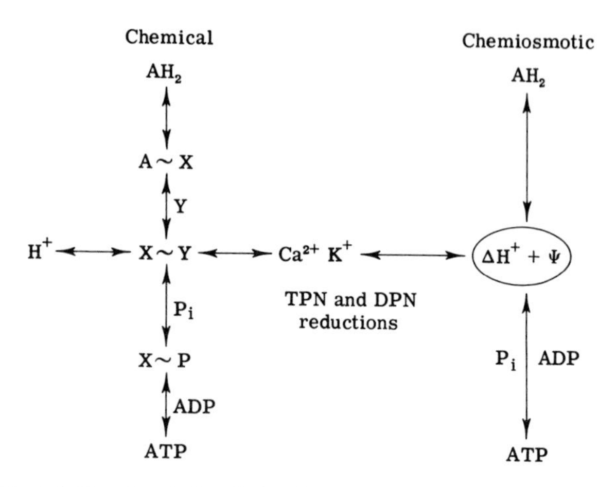

Fig. 3-3 The chemical and chemiosmotic hypotheses of oxidative phosphorylation.

3-3). In intact mitochondria, energy-dependent reactions such as active transport of Ca^{2+}, Mg^{2+}, or K^+ can be driven by substrate oxidation in the presence of oligomycin which prevents ATP formation. Similar experiments on two energy-dependent reductions can be performed with submitochondrial particles: (a) the reduction of DPN by succinate and (b) the energy-dependent transhydrogenation between TPN and DPNH. These reactions can be driven by either oxidation energy or by ATP but only in the latter case are the reactions sensitive to oligomycin. Submitochondrial particles that are deficient in coupling factors also catalyze the reduction of DPN by succinate but only if driven by the oxidation energy (e.g., at the third site). Energy transfer inhibitors such as oligomycin not only fail to inhibit these processes, but may stimulate or even be required. These effects of energy transfer inhibitors and the finding (Lee and Ernster, 1966) that oligomycin at low concentrations stimulates oxidative phosphorylation suggest that this inhibitor has an effect on the maintenance of the high-energy state.

It is rather unfortunate that the formation of a high-energy state in the absence of P_i, a well-documented fact, has been ignored by some outstanding chemists who have dreamed up ingenious, but worthless, models of oxidative phosphorylation which postulate the entrance of P_i prior to or simultaneous with oxidation.

According to the chemical hypothesis, the transport systems of Ca^{2+}, K^+, H^+ are on a side path driven by a nonphosphorylated intermediate in a reversible manner as shown in Fig. 3-3. This means that X~Y and ATP could be generated by a gradient of either protons or of cations by reversal of the direction of flow. In fact, ATP formation has been demonstrated on dissipation of a proton gradient in chloroplasts (Jagendorf and Uribe, 1966) and of a potassium gradient in mitochondria (Cockrell *et al.,* 1967).

What is the evidence for A~X and X~P, the two other high-energy intermediates of oxidative phosphorylation? It is only fair to say that the evidence for either is neither convincing nor generally accepted. Evidence for A~X derived from spectral changes in energized membranes is far from conclusive and subject to alternative interpretations. Evidence for X~P, which I personally view more sympathetically, is also indirect and is generally ignored. Both viewpoints will be discussed later in greater detail, but it should be emphasized that the existence of an A~X intermediate is inconsistent with the chemiosmotic hypothesis which does not allow for a direct interaction of A, a member of the respiratory chain, and X, a member of the coupling device. On the other hand, an intermediate X~P could be incorporated into the chemiosmotic hypothesis without sacrifice of the basic principle.

The Chemiosmotic Hypothesis

The most important and unique feature of Mitchell's hypothesis is that the primary event catalyzed by the oxidation chain is the formation of a proton motive force which is the sum of a membrane potential and a ΔpH. The protons are then driven back via a proton pump, the oligomycin-sensitive ATPase complex of the membrane (Fig. 3-4). During this process ATP is formed. This formulation includes two important basic assumptions: a com-

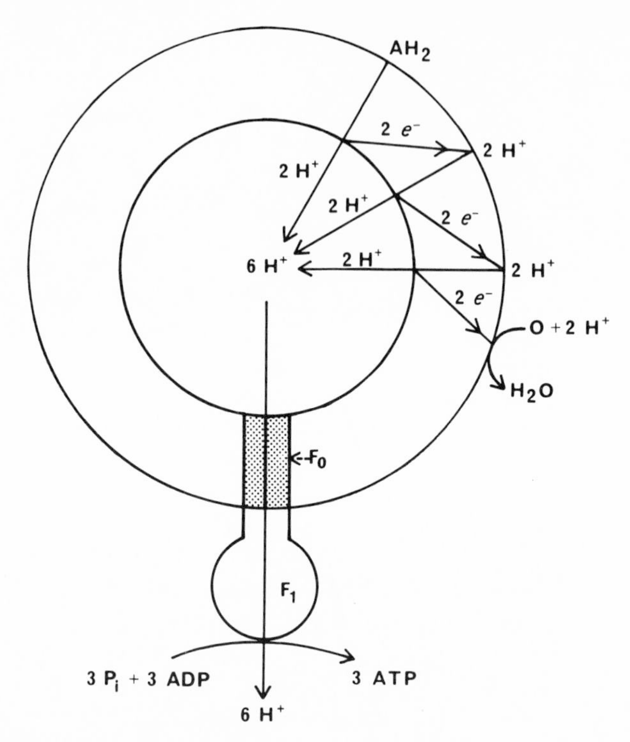

Fig. 3-4 The chemiosmotic reactions; the loops of the oxidation chain and the ATP-driven proton pump.

partment is required and the mitochondrial membrane is impermeable to protons.

According to Mitchell the formation of a proton motive force is achieved by an asymmetric organization of the oxidation chain in three loops. Mitchell (1966) has formulated in great detail the individual components of these loops. I wish that proponents of new ideas would not feel compelled to fill in all the loop holes (or loops) of their hypothesis. It is curious though, how little attention was paid to Mitchell's hypothesis of 1961 when he proposed it in terms of basic principles, which are as good today as they were in 1961. It is only when in 1966 Mitchell included detailed features that could be attacked that his hypothesis was widely reviewed.

Lesson 9: If you propose a brilliant, new hypothesis and avoid being specific, you probably will be ignored. If you want attention be sure to include a few features that are not essential for the main theme, but open to attack.

The operation of the oxidation loops at sites 1 and 3 is illustrated in Fig. 3-5. At the first site $DPNH_2$ donates at the matrix side (M-side) of the membrane 2 hydrogens to a hydrogen carrier which releases 2 protons at the other side (C-side) of the membrane, while the electrons are channeled back via electron carriers to the M-side. There they combine with a proton from the medium and the second hydrogen carrier repeats the process in a second loop, etc.

According to Mitchell's formulation of 1966 the operation of the coupling device, the oligomycin-sensitive proton pump, takes place via intermediates XO^- and Y^- which respond to the membrane potential by moving away from the negatively charged side. There is no experimental evidence for the existence of these negatively charged anions and I mention this formulation only

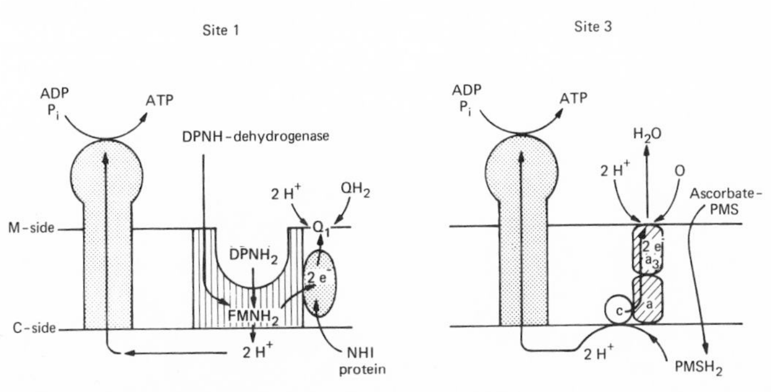

Fig. 3-5 The chemiosmotic formulation of the operation of site 1 and site 3 of oxidative phosphorylation.

because it is an ingenious scheme which shows how a membrane potential can participate in a chemical reaction. In the first and latest formulation of the chemiosmotic hypothesis X~Y, the high-energy state, is identical with the proton motive force.

According to the chemiosmotic hypothesis the essential feature of the coupling device (oligomycin-sensitive ATPase) is that it is a proton pump functioning in reverse, i.e., utilizing a proton flux to generate ATP. As will be discussed later, other ion pumps, e.g., the Ca^{2+} and the Na^{+} pumps, are also reversible and can utilize ion gradients to generate ATP. Thus Mitchell's formulation of the basic principle of ATP formation by an ion gradient has ample experimental support from other systems.

The "Conformational Hypothesis"

It was first proposed by Boyer (1965) and King *et al.*, (1965) that energy conservation may take place by conformational changes in electron transport enzymes. This formulation is clearly a variant of the chemical hypothesis substituting A* for A~X as shown in Fig. 3-6. More recently Boyer (1974) has proposed another "conformational hypothesis" that centers on the mode of action of the mitochondrial ATPase. The key feature is that a conformational change of the protein allows the release of tightly bound ATP from the ATPase, a step that requires energy. In fact he suggests that little if any energy is required for the actual formation of ATP from ADP and P_i at the hydrophobic site of the catalytic center of the enzyme. This attractive idea is not really a hypothesis of oxidative phosphorylation as such, since it does not include a mechanism of energy transfer from oxidation to phosphorylation. If the conformational change is transmitted directly from components of the electron transport chain to the ATPase, we are dealing with a chemical hypothesis; if conformational changes are induced in response to proton translocation, we are dealing with a chemiosmotic mechanism.

Conformational changes in a protein must take place during catalysis. I should like to mention here that Mitchell suggested to me in 1965 that we should look for ATP formation by imposing a

Chemical	Conformational
$AH_2 + B + X \rightleftharpoons A{\sim}X + BH_2$	$AH_2 + B \rightleftharpoons A^* + BH_2$
$A{\sim}X + Y \rightleftharpoons X{\sim}Y + A$	$A^* + X \rightleftharpoons A + X^*$
$X{\sim}Y + P_i + ADP \rightleftharpoons X + Y + ATP$	$X^* + P_i + ADP \rightleftharpoons X + ATP$

Fig. 3-6 Similarity between the chemical and the (original) conformational hypotheses of oxidative phosphorylation. Note: * is substituted for Y; there is still interaction between A and X.

pH change on the soluble mitochondrial ATPase. Although such experiments have been consistently negative with either F_1 or CF_1, net ATP formation was observed with stoichiometric amounts of the Ca^{2+}-ATPase of sarcoplasmic reticulum (Knowles and Racker, 1975a). In any case ion-induced conformational changes of the protein must play a role in ATP formation. We shall return to these problems in Lecture 7 dealing with the mechanism of ion pumps.

Rather than enlarging on the theoretical arguments in favor or against one or the other hypothesis, I shall spend the rest of this lecture analyzing certain experimental approaches that have bearing on the mechanism of oxidative phosphorylation.

Experimental Approaches

The Role of the Membrane and Requirement for a Compartment

According to the chemical hypothesis oxidative phosphorylation could be catalyzed by a flat piece of membrane or even by solubilized enzymes. In contrast, the chemiosmotic hypothesis requires a compartment that permits the formation of a proton motive force. In this respect the chemiosmotic hypothesis is much more vulnerable and could be eliminated by the demonstration of oxidative phosphorylation in solution. Even the uncoupler-sensitive $^{32}P_i$–ATP exchange reaction catalyzed by the transmembranous uncoupler- and oligomycin-sensitive ATPase cannot function in solution according to the chemiosmotic hypothesis. Yet, claims have been made that an uncoupler-sensitive ATP synthetase has been solubilized (Fisher *et al.,* 1971a). It is now generally agreed that small vesicles of the type we have described (Arion and Racker, 1970) are responsible for this uncoupler-sensitive exchange. The uncoupler-insensitive exchange present in the same preparation is apparently not related to oxidative phosphorylation since it is sensitive to avidin (You and Hatefi, 1973).

There have been several other claims of solubilized systems of oxidative phosphorylation, but until now none has survived careful scrutiny or attempted repetitions in other laboratories.

Evidence from reconstituted systems suggests that oxidative phosphorylation requires the presence of phospholipid vesicles that can form a compartment. Also, the experiments on the structural role of F_1 and CF_1 which I discussed in the first two lectures are more readily understood in terms of the Mitchell hypothesis which requires a closed compartment.

Mechanism of Action of Uncouplers and Ionophores

According to the chemical hypothesis the proposed function of uncouplers is to hydrolyze X~Y. The large range of chemicals with divergent structures made this an unattractive explanation. Mitchell proposed instead that uncouplers act as lipid-soluble weak acids which transport protons across the membrane and thereby collapse the proton motive force. This formulation received considerable support from later studies on phospholipid model membranes (Bielawski *et al.,* 1966; Mueller and Rudin, 1969; Skulachev, 1971). Quantitative differences have been observed when the responses of these model membranes to various uncouplers were compared with those of the mitochondrial membrane. However, this is hardly surprising in view of the vast differences in chemical constitution and morphological architecture of the two systems under comparison.

The mode of action of uncouplers as proton ionophores could be accommodated within the chemical hypothesis. A proton pump linked in parallel rather than in series with the coupling device of oxidative phosphorylation could account for uncoupling by proton ionophores in a way similar to the uncoupling of mitochondrial phosphorylation by valinomycin in the presence of potassium. On the other hand, the evidence that the proton pump of the mitochondrial ATPase is in fact the coupling device of oxidative phosphorylation is rather compelling, particularly in view of reconstitution experiments that will be discussed later.

The Proton Motive Force

It was a major contribution of Mitchell to have drawn the attention to the movement of protons during electron transport in oxidative and photophosphorylation (Mitchell, 1966; Hinkle and Horstman, 1971). Jagendorf and Uribe (1966) have shown that in chloroplasts a pH gradient can give rise to ATP formation in the dark. Since CF_1 is required for this process (McCarty and Racker, 1966) it appears that the reaction is channeled via the coupling device of photophosphorylation.

It is now well established that a pH gradient is actually formed in chloroplasts and a ΔpH up to 3.5 pH units was determined (Rottenberg *et al.,* 1972; Portis and McCarty, 1974). Moreover, there appears to be an excellent correlation between the ΔpH and the rates of photophosphorylation. Although a contribution to ATP formation by a membrane potential was not observed when the ΔpH was large, such a contribution was observed when the pH gradient was small (Avron *et al.,* 1973).

Recently, Thayer and Hinkle (1975) have demonstrated ATP generation in submitochondrial particles induced by a combination of a pH gradient and a membrane potential. Their finding that the initial velocity of this ATP generation is actually faster than that coupled to DPNH oxidation is critical to the question of the primary role of proton translocation. Whether the proton motive force that can be calculated from these experiments is sufficient to satisfy thermodynamic considerations is still being disputed. As pointed out in Lecture 1, such thermodynamic calculations are treacherous since they include assumptions that may be erroneous. For example the assumption of a H^+/ATP ratio of 2 has recently been challenged at least in the case of chloroplasts. If a ratio of 4 is in fact the correct value (Rumberg and Schröder 1973), all thermodynamic objections would be resolved.

The Membrane Potential

If anyone wants to take the direct measurements of the membrane potential in mitochondria by Tupper and Tedeshi (1969a,b,c) at their face value, the formation of a membrane potential during respiration has been disproved. Since I am not accepting this judgment you might say that I am ignoring facts.

> *Lesson 10:* You have heard the famous quotation by Thomas Henry Huxley of "The Tragedy of Science: a beautiful hypothesis slain by an ugly little fact." The trouble with this pronouncement is that it lacks a definition of the "ugly little fact." I have two objections against this famous citation. I don't like the aspect of the beauty contest. The beauty of a hypothesis that is wrong is illusionary. If the "ugly little fact" has cured us of an illusion, it is neither ugly nor little. But a study of the history of science teaches us that many beautiful hypotheses have survived hundreds of so-called "ugly little facts," because they really were ugly and little. Sometimes they could be explained by minor modification of the hypothesis, sometimes they were not facts but artifacts. Before pronouncing a verdict of murder of a hypothesis, we should make certain that we have a corpse. A good hypothesis can survive many ugly little facts and is worth a few hundred negative experiments.

Let us examine the case of the membrane potential. Into a giant mitochondrion of *Drosophila* (about 4 μm in diameter) an electrode was inserted. The insertion of an electrode into a mitochondrion, even into a giant mitochondrion, is a difficult operation when you consider the relative dimensions. It is like pushing a baseball bat into a cat. Anyhow, instead of a large negative potential as predicted by the chemiosmotic hypothesis, a small positive potential was recorded. A very low resistance was also noted which is difficult to understand in the light of the ion impermeability of the membranes. Actually, it is not at all certain that the electrode penetrated into the matrix of the mitochondria. The mitochondrial membranes are resiliant and contain numerous foldings. It is conceivable that the electrode was placed between the outer and inner membrane.

On the other hand, there is indirect evidence for a membrane potential in mitochondria (Mitchell and Moyle, 1969; Skulachev, 1971). Synthetic lipid-soluble ionized compounds such as tetraphenylboron or picrate penetrate the membrane depending on the charge. Anions move out of mitochondria, cations move in. In submitochondrial particles, which are "inside-out," anions move in, cations move out. Energy for these movements can be supplied by either oxidative energy or by ATP or by a membrane potential generated nonenzymatically. Skulachev (1971) has shown that phenyl dicarbaundecaboron (PCB) is a hydrophobic anion that changes its distribution in the membrane with the appearance of an electric potential difference. In mitochondria respiration results in an efflux of PCB, and in submitochondrial particles PCB is taken up. In both cases the movements are prevented by rotenone, an inhibitor of the respiratory chain. Synthetic cations moved in the opposite direction. The same movements could be observed when an ion gradient (K^+ plus valinomycin) was used instead of respiration to create a membrane potential.

Partial Reactions of Oxidative Phosphorylation

There are numerous partial reactions that have been used for studies of the mechanism of oxidative phosphorylation. They will be discussed in detail in the next two lectures.

Intermediates

Mitchell (1966) has used the fact that, in spite of numerous attempts no intermediates of oxidative phosphorylation have been found, as an argument against the chemical hypothesis. I cannot share this view. If biochemists were not smart enough to think of the chemiosmotic hypothesis before 1961 perhaps they are not smart enough to isolate an intermediate.

Studies of the Na^+-K^+ pump of the plasma membrane and of the Ca^{2+} pump of sarcoplasmic reticulum have revealed the existence of phosphorylated intermediates which I shall discuss in Lecture 6. Since these pumps can also generate ATP by dissipation of an ion gradient, the analogy with the mitochondrial proton pump seems obvious. I therefore feel that a search for a chemical phosphorylated intermediate in oxidative phosphorylation should be continued. Experiences with the Na^+ and Ca^{2+} pump help to explain why previous attempts have met with failure and why indeed it may be difficult to demonstrate such an intermediate of F_1 in the future. However, it must be emphasized that (a) the chemiosmotic hypothesis can survive the inclusion of such an intermediate and (b) a phosphorylated intermediate has not as yet been identified in the operation of the mitochondrial proton pump.

Chance and Schoener (1966) reported that ATP induces uncoupler-sensitive changes in the spectral properties of cytochrome *b*. Slater *et al.* (1970) confirmed spectral changes caused by ATP though at somewhat different wavelengths. Wilson and Dutton (1970) have reported that ATP lowers the midpoint potential of cytochrome a_3. These and similar effects of ATP on the other sites of oxidative phosphorylation have been interpreted in terms of high-energy intermediates of respiratory chain components. However, Hinkle and Mitchell (1970) who observed changes in the midpoint potential of cytochrome *a* as well as a_3 point out that these findings were predicted on the basis of the expected positive charge generated during oxidation at the C-side of the membrane, which should lower the apparent redox potential of the cytochromes by a "pulling effect" on the electrons from cytochrome a_3 (near the inner surface) to cytochrome c_1 (near the outer surface). One must admire the courage of such a prediction which was made before the location of cytochrome c_1 was actually known.

Caswell (1971) has criticized the work of Wilson and his collaborators and has stressed some of the complications of measuring midpoint potential in the presence of artificial electron mediators. Slater concludes (1972) "it is possible to explain the effects of ATP on the apparent redox potential of cytochrome *b* . . . without assuming a direct effect on the cytochrome *b* molecule itself. This is a disappointing conclusion." He also says, "It appears likely that the ATP-induced red shift is associated with an increase of the primary energy pressure rather than with secondary reactions. A strong indication in this direction is the finding that addition of valinomycin, which energizes the particles by reversal of the energy-requiring K^+ uptake, has the same effect as energization by ATP." I would not be surprised if Mitchell would quote the same experiment as evidence that the spectral shifts are secondary to the formation of a membrane potential and analogous to the changes induced by ATP, e.g., on the fluorescence of anilinonaphthalene sulfonate.

Although the search for intermediates has not provided generally accepted conclusions, the spectral analyses have been fruitful in the recognition of different species of the electron transport chain and of the order of interaction of the respiratory catalysts in the process of electron transport.

(non-biological)

Lecture 4

The Coupling Device

The little I know, I owe to my ignorance.
Sacha Guitry

I love to doubt as well as know.
Dante, *The Inferno*

The process of oxidative phosphorylation is catalyzed by a multienzyme system embedded in the inner mitochondrial membrane. It consists of a multienzyme system referred to as the respiratory or oxidation chain which catalyzes the hydrogen transfer from substrates (DPNH, succinate, ascorbate) to molecular oxygen as the final electron and proton acceptor. It also contains a coupling device which catalyzes the formation of ATP from ADP and P_i during oxidation. The oxidation chain and the coupling device are tightly coupled in intact mitochondria, i.e., oxidation does not take place unless ADP and P_i are available for the formation of ATP.

The experimental approaches to the problem of oxidative phosphorylation can be divided into two categories: (1) examination of partial reactions and intermediates and (2) resolution and reconstitution of the catalysts. In this lecture I shall discuss the coupling device and its partial reactions; the next lecture will be concerned with the oxidation chain.

Partial Reactions and Components of the Coupling Device

Exchange Reactions

Investigations of exchange reactions with isotopes have helped greatly in the elucidation of the mechanism of oxidative phosphorylation. Incorporation of ^{18}O from water into P_i or ATP or of radioactive P_i or ADP into ATP will be referred to as H_2O–P_i, P_i–ATP, etc., exchanges. I shall not discuss in detail the earlier work which I have reviewed previously (cf. Racker, 1965).

Experiments with ^{18}O were instrumental in establishing that P_i interacts with the coupling device before ADP and that the bridge oxygen between the terminal and second phosphorus in ATP is derived from ADP (Boyer, 1958). Studies on the P_i–ATP exchange established that the coupling device can operate in membranes which do not contain an oxidation chain (Groot *et al.*, 1971; Kagawa and Racker, 1971). The oligomycin-sensitive ADP–ATP exchange (Wadkins and Lehninger, 1963) as well as the H_2O exchanges (Hinkle *et al.*, 1967a) provided indirect but suggestive evidence for a high-energy phosphorylated intermediate. The existence of a nonphosphorylated high-energy intermediate or state (X~Y) was established by studies of energy-dependent partial reactions that took place in the absence of ATP, e.g., reversal of oxidation and the transhydrogenase reaction (cf. Ernster and Lee, 1964). In addition to information on the mechanisms of oxidative phosphorylation, partial reactions have been very useful for the assay of membrane components during the resolution and reconstitution of the coupling device.

According to Mitchell (1966) the coupling device is an ATP-driven proton pump operating in reverse. In the simplest formulation of the chemiosmotic hypothesis X~Y is identical with the proton motive force. In the more complex version (Mitchell, 1966) X~Y is an intermediate in equilibrium with the proton motive force.

The formulation of the operation of the coupling device shown in Table 4-1 is consistent with either the chemiosmotic or chemical hypothesis of oxidative phosphorylation. It is not sup-

TABLE 4-1 **Hypothetical Scheme for Operation of Coupling Device in Ion Pumps**

$$\text{Step I.}\quad \left[R-C\begin{matrix}\nearrow O\\ \searrow OH\end{matrix}\right] + H_3PO_4 \rightleftharpoons R-C\begin{matrix}\nearrow O\\ \searrow OPO_3H_2\end{matrix} + H_2O$$

$$\text{Step II.}\quad R-C\begin{matrix}\nearrow O\\ \searrow OPO_3H_2\end{matrix} + ADP \rightleftharpoons R-C\begin{matrix}\nearrow O\\ \searrow OH\end{matrix} + ATP$$

ported by strong evidence of experiments with mitochondria, but is derived from work with other ion pumps which will be discussed in Lecture 6. The purpose of proposing this mechanism here is for the sake of discussing partial reactions in more concrete terms and for the design of future experiments. In the formulation shown here an activated carboxyl group interacts with inorganic phosphate. Alternatively an activated P_i could react with a carboxyl group or even directly with ADP. The latter formulation is favored by Mitchell (1974). It does not however lend itself readily for the interpretation of the H_2O–ATP exchange in the absence of reversal of the overall reaction.

P_i–ATP Exchange

This reaction is dependent on the presence of vesicular structures (Kagawa and Racker, 1971). The reaction is representative of the operation of the complete coupling device but is actually more complex than a partial forward reaction of oxidative phosphorylation. It includes the reversal of step I and step II (Table 4-1) as well as the complete forward reaction from $^{32}P_i$ to ATP. In the chemiosmotic formulation it requires the translocation of protons via the transmembranous ATPase in both directions. Since the relative activities of the forward and reverse reaction may vary greatly with different conditions and preparations, as most strikingly illustrated in chloroplasts, it is not surprising that this partial reaction of oxidative phosphorylation does not always conform to the properties of the overall system.

ADP–ATP Exchange

This reaction should be most useful for the study of the postulated phosphorylated intermediate XO~P. Unfortunately there are complications that have restricted the usefulness of this reaction. In mitochondria the ADP–ATP exchange is sensitive to oligomycin and uncouplers (Wadkins and Lehninger, 1963), but submitochondrial particles which catalyze oxidative phosphorylation do not appear to catalyze the ADP–ATP exchange (Zalkin *et al.,* 1965). However, it was pointed out that experimental difficulties are associated with the measurements.

H_2O–ATP and H_2O–P_i Exchanges

The H_2O exchanges are important partial reactions of the coupling device and at present provide the only indirect evidence for a phosphorylated intermediate of submitochondrial particles. It was proposed many years ago (Cohn and Drysdale, 1955) that there are two sites of water entry during oxidative phosphorylation: one into P_i and one into ATP. In early work (Cohn, 1953) a stimulation of the H_2O–P_i exchange by adenine nucleotides and later a complete dependence was observed (Hinkle *et al.,* 1967a; Mitchell *et al.,* 1967). The formulation shown in Table 4-1 with a phosphorylated intermediate, which could contain a pentavalent phosphorus as suggested by Cohn (1958) and by Haake and Westheimer (1961), accounts for the H_2O–ATP exchange as well as for a nucleotide dependent H_2O–P_i exchange (Racker, 1970b). The marked stimulation of the H_2O–ATP exchange by P_i (Hinkle *et al.,* 1967a) can also be explained by this scheme, since P_i would be required to maintain a high steady-state level of $XO{\sim}PO_3H_2$. Although a concerted mechanism (Boyer, 1968) could account for the adenine nucleotide requirement of the H_2O–P_i exchange, it fails to explain the occurrence of the rapid H_2O–ATP exchange in chloroplasts in which the overall reversal is virtually absent (Avron *et al.,* 1965; Shavit *et al.,* 1967). It was recently proposed (Boyer, 1974; Slater, 1974) that firmly bound ATP may be the long sought after phosphorylated high-energy intermediate. The O~P bond of protein-bound ATP of membrane-bound F_1 may be labilized to account for the H_2O–ATP exchange. However, it should be noted that soluble F_1 which contains firmly bound ATP does not catalyze this exchange. Before the significance of bound ATP can be evaluated it needs to be shown that bound ATP turns over very rapidly, consistent with the rate of phosphorylation. It should thus be possible to establish kinetically whether bound ATP could in fact function as an intermediate. Alternatively, bound ATP may have a regulatory function or may serve as a stabilizer in the interaction between subunits of the ATPase as was shown in the experiments on the reactivation of dissociated CF_1 (S. Lien and Racker, 1971).

The Oligomycin-Sensitive Hydrolysis of ATP

This reaction is probably the simplest representative of the coupling device. It is the only partial reaction that does not appear to require closed vesicles. However, it is dependent on phospholipids, multiple coupling factors, and the hydrophobic components of the ATPase complex.

The Coupling Factors

Proteins that can be removed from the mitochondrial membrane and are required for the reconstitution of a functional coupling device have been called coupling factors. We prefer to retain this name for the time being because it is operational and noncommittal and because the exact function of some of these factors is still uncertain. Adoption of a name such as ATP synthetase for a mixture of soluble coupling factors (Fisher *et al.*, 1971a) is not advisable unless net synthesis of ATP can be demonstrated.

The concept of "coupling factors" was challenged by the discovery by Lee and Ernster (1966) that low concentrations of oligomycin (0.2 to 0.4 μg/mg protein) stimulated phosphorylation in A-particles (= EDTA particles) that were deficient in coupling factors. This observation led to the discovery that coupling factors can have a catalytic and a structural function (Racker, 1967) and that oligomycin can substitute only for the latter.

As in the case of the vitamins the early history of coupling factors is very confusing. Such confusion and an overkill in nomenclature are not surprising in the analysis of complex multicomponent systems when many investigators are involved using different materials and different assays. Four coupling factors from mitochondria have been isolated and characterized sufficiently to warrant discussion: F_1, F_2, OSCP, and F_6. In most partly resolved submitochondrial particles (though not in all) OSCP replaced F_3 and F_5 (Fessenden-Raden *et al.*, 1969). F_4 is a mixture of F_2, OSCP, and F_6. Factor A (Andreoli *et al.*, 1965) and factor B (Lam *et al.*, 1967; Racker *et al.*, 1970c) are other names for F_1 and F_2, respectively. Although the oligomycin-sensitive ATPase does not seem to require F_2, discussion of this protein is included here for the sake of continuity.

Mitochondrial Coupling Factor 1 (F_1). F_1 (= mitochondrial ATPase) (Pullman *et al.*, 1960) is required for oxidative phosphorylation (Penefsky *et al.*, 1960) and for all partial reactions of the coupling device (cf. Racker, 1970b). The protein is located on the matrix side of the inner mitochondrial membrane where it appears as spheres of about 85 Å in diameter attached by a stalk. The identity of the stalk has still not been established. It may be OSCP, F_6, or a subunit of F_1 which is extended into a stalk when the protein interacts with the membrane. The physical relationship of the chloroplast coupling factor CF_1 with the membrane, which I shall discuss shortly, applies also to F_1. All the evidence now suggests that the appearance of the inner membrane spheres of mitochondria is not an artifact of the staining procedure.

The physical chemical properties of F_1 have been extensively reviewed (Senior, 1973). In most respects such as subunit composition, molecular weight, cold lability, and amino acid composition, it is remarkably similar to the spinach coupling factor CF_1 (cf. Lecture 2) and to the *E. coli* ATPase (Nelson *et al.*, 1973; Futai *et al.*, 1974).

The ATPase Activity of F_1 and Its Inhibition by a Mitochondrial Protein. It never fails to confuse students that a coupling factor that is involved in ATP formation has ATPase activity which appears to defeat the purpose of its function. Ever since we discovered ATPase activity in soluble F_1, we have emphasized that this activity is an artifact. We proposed (Penefsky *et al.*, 1960) that the membrane-bound enzyme is a phosphotransfer enzyme which catalyzes the formation of ATP from ADP and X~OP, a phosphorylated intermediate (see Table 4-1, step II). Many phosphotransfer enzymes have been shown to interact with hydroxyl groups other than those of the natural substrates, and some may recognize the hydroxyl group of water and thus function as an ATPase. In some cases reactivity with water appears only after aging or mistreatment of the enzyme as in the case of the phosphatase activity of glyceraldehyde-3-phosphate dehydrogenase (cf. Racker, 1965).

> *Lesson 11:* You can fool some of the enzymes most of the time, you can fool most of the enzymes some of the time, but you cannot fool all of the enzymes all of the time. We can fool many enzymes particularly after mistreating them thus permitting us to study them under unusual circumstances. But can we expect them to behave normally when we poison them with cacodylate buffer or put them to sleep with veronal buffer? Perhaps in the end the enzyme is fooling us.

The hydrolytic activity of F_1 in intact mitochondria is very small and it becomes manifest only when the inhibitor protein (Pullman and Monroy, 1963) is dissociated. In chloroplasts the binding of the inhibitor is so tight (Nelson *et al.*, 1972b) that there is no detectable ATPase activity unless the chloroplasts are mistreated, e.g., with trypsin or heat. We therefore look upon the inhibitor protein as a regulatory subunit of the coupling factor.

The difference in the dissociation of the ATPase–inhibitor complex in mitochondria and chloroplasts may well have physiological significance, since in mitochondria the dissociation appears to be controlled by the energy state of the membrane (Van de Stadt *et al.*, 1973).

The Chloroplast Coupling Factor (CF_1). CF_1 has a molecular weight of about 350,000 (Farron, 1970). In sodium dodecyl sulfate it dissociates into 5 subunits (Racker *et al.*, 1972). There is experimental evidence that some of the subunits of CF_1 have a role in catalysis, and others in the attachment to the membrane or in the regulation of the ATPase activity. The smallest of the subunits, ϵ (13,500 MW), was identified as a regulatory rather than a catalytic component of the enzyme (Nelson *et al.*, 1972b). The isolated pure ϵ subunit inhibits the ATPase activity of CF_1 without interfering with its function as a coupling factor. Thus it acts exactly like the mitochondrial protein inhibitor of ATPase (Pullman and Monroy, 1963). However, the spinach inhibitor is much more hydrophobic and firmly bound to CF_1 so that it cannot be dissociated by the procedure used for the isolation of the mitochondrial inhibitor.

The subunits of CF_1 were isolated in pure form and injected into rabbits for antibody production (Nelson *et al.*, 1973). The effect of these antibodies on the ATPase activity of CF_1 is shown in Table 4-2. The antibodies against the largest subunit α (59,000 MW) and against the second largest β (56,000 MW) agglutinated chloroplasts; the antibodies against the smaller subunits did not. ATP hydrolysis was not affected by any single subunit antibody but was inhibited by a combination of anti-α and the antibody against the γ subunit (37,000 MW). The most potent inhibitor against photophosphorylation was the antibody against the γ subunit (Table 4-3). At somewhat higher concentrations anti-α also inhibited photophosphorylation. The divalent antibody against the β subunit was ineffective as inhibitor; however, experiments (Gregory and Racker, 1973) with monovalent anti-β revealed significant inhibition of photophosphorylation although (as expected) agglutination was no longer observed. It is quite apparent that agglutination of the chloroplasts does not play a role in the inhibition of photophosphorylation. Thus far we have

TABLE 4-2 **Inhibition of Ca^{2+}-ATPase of CF_1 by Antisera against CF_1 Subunits**[a]

	ATPase activity (μmoles P_i/min/mg protein)
Control	26.5
20 μl anti-α	18.0
10 μl anti-α + 10 μl anti-β	21.2
10 μl anti-α + 10 μl anti-γ	7.3
10 μl anti-α + 10 μl anti-δ	21.7
20 μl anti-β	20.3
10 μl anti-β + 10 μl anti-γ	19.7
10 μl anti-β + 10 μl anti-δ	21.2
20 μl anti-γ	25.7
10 μl anti-γ + 10 μl anti-δ	26.4
20 μl anti-δ	19.9

[a]Experimental conditions were as described by Nelson *et al.* (1973).

TABLE 4-3 **Inhibition of Photophosphorylation by Antisera against CF_1 Subunits**[a]

Addition of antibody		PMS cyclic photophosphorylation (μmoles/hour/mg chlorophyll)
None		1010
Control serum	30 μl	1045
Anti-β	30 μl	1025
Anti-γ	5 μl	238
	10 μl	69
Anti-δ	30 μl	1140
Anti-α	10 μl	557
	30 μl	252

[a]Experimental conditions were as described by Nelson *et al.* (1973).

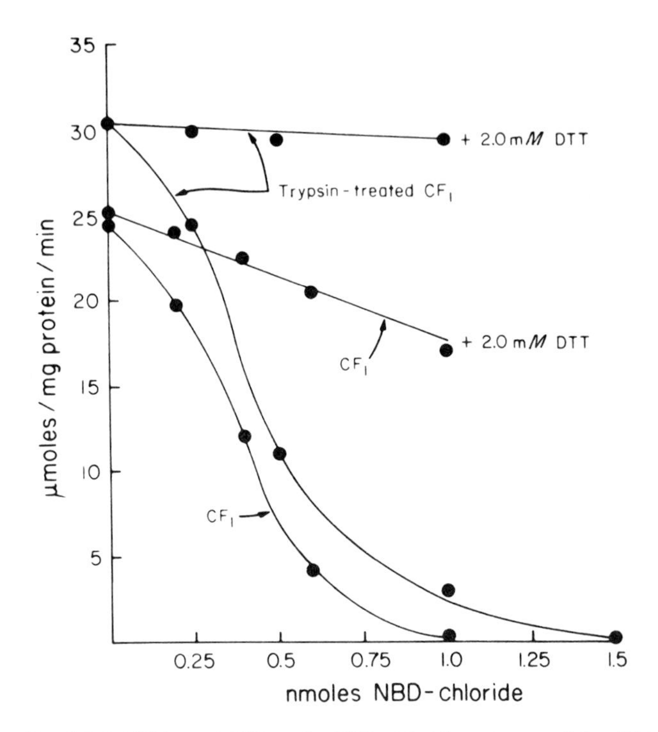

Fig. 4-1 Inhibition of ATPase by NBD-chloride and reversal by dithiothreitol. Experimental conditions were as described by Deters *et al.* (1975). CF_1 and trypsin-treated CF_1 were incubated with the indicated amounts of NBD-chloride overnight at room temperature and then tested for ATPase activity in the absence and presence of DTT.

not observed inhibition of photophosphorylation by the antibody against the δ subunit.

As shown in Fig. 4-1, NBD-chloride inhibits the ATPase activity of CF_1 at very low concentrations. The inhibition is reversed by dithiothreitol (Deters *et al.*, 1975). It was shown with radioactive NBD-chloride that the inhibitor interacts specifically with the β subunit. NBD-Chloride also interacts with the β subunit of F_1 (Ferguson *et al.*, 1975a,b). One equivalent is sufficient to inactivate both ATPase and coupling factor activity. The γ subunit of CF_1 was implicated in function by the experiments of

McCarty and Fagan (1973) who have shown that *N*-ethylmaleimide interacts with this subunit specifically during illumination, thereby inhibiting photophosphorylation. Removal of the δ subunit by heat treatment in the presence of digitonin followed by passage through a DEAE column (Deters *et al.*, 1975) yielded an ATPase which contained the α, β, and γ subunits but had no coupling activity. In contrast, heated CF_1 preparations that had not been passed through a column retained their coupling activity, presumably due to reconstitution of the δ subunit with the heated protein during interaction with the chloroplast membrane. In the case of *E. coli*, an ATPase preparation missing the δ subunit lacked coupling factor activity (Bragg *et al.*, 1973) and reconstitution of coupling factor activity by addition of purified δ subunits was recently demonstrated (Smith and Sternweis, 1975). It was shown that the δ subunit is required for the attachment of the coupling factor to the membrane.

A plant bioflavonoid called quercetin inhibits many membranous ATPases, including the ATPase activity of CF_1 (Deters *et al.*, 1975). After treatment with trypsin, CF_1 contained only α and β subunits with undiminished ATPase activity. The ATPase was still sensitive to quercetin but had become resistant to inhibition by the ϵ subunit. In other respects quercetin acted very similarly to the ϵ subunit of CF_1. I shall return to quercetin in Lecture 8 on the energy metabolism of tumor cells.

Topography of CF_1. The observations on the interaction of CF_1 with various subunit antibodies have led us to a tentative formulation of the topography of CF_1 and the relationship of its subunits to the membrane (Fig. 4-2). Although such a formulation is obviously speculative, it allows us to design experiments on the reconstitution of CF_1 from its subunits. The two large subunits α and β appear to be readily accessible to the corresponding antibodies which agglutinate chloroplasts. The γ subunit which interacts with its antibody as shown by the potent inhibition of phosphorylation should be in a position which is sterically not permissive for the formation of a lattice required for agglutination. We have therefore tentatively positioned the γ subunit close to the membrane, overshadowed by one of the larger

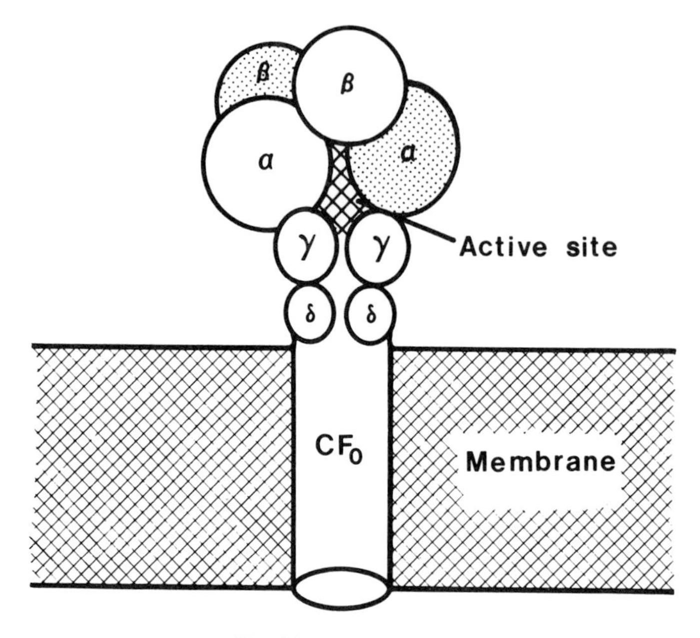

Fig. 4-2 Topography of CF_1.

subunits, but it could be located in another position with limited accessibility to the antibody. The accessibility of the δ subunit is also limited, since δ antibodies do not agglutinate chloroplasts. Proximity of the δ subunit to the membrane is indicated by the above mentioned experiment with *E. coli* F_1 and CF_1. Moreover, removal of the small subunits of CF_1 by trypsin resulted in a loss of binding of the protein to the membrane (Deters *et al.*, 1975). Thus the δ and perhaps also the γ subunit may contribute to the stalk seen in electron micrographs. The ε subunit which has not been inserted into the Fig. 4-2 should be located close to the active center and in the vicinity of the γ subunit since the antibody against the γ subunits interfered with the interaction of CF_1 with the ε subunit. Two points should be emphasized with respect to these tentative conclusions which are largely based on the studies with antibodies. First, conformational changes induced by antibodies may affect the reactivity of the protein at a

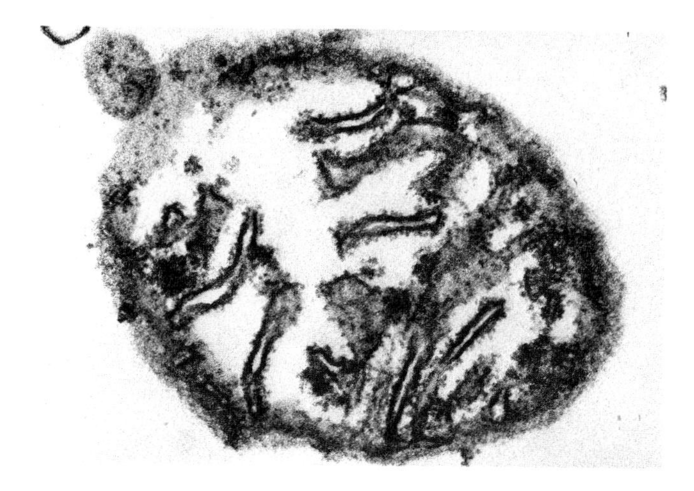

Fig. 4-3 Electron micrograph of mitochondria fixed with glutaraldehyde and stained with methanolic osmium tetroxide after sectioning (magnification × 56,000). Experimental conditions were as described by Telford and Racker (1973).

distance removed from the site of antibody interaction. Such changes could invalidate the conclusions. Second, the assignment of two copies for each of the subunits is also less than certain. Although we believe that in the case of the α and β subunits two is still the most plausible number, in the case of the minor subunits the number may be either smaller or larger. The value of such a tentative scheme as shown in Fig. 4-2 rests with its usefulness for the design of experiments. If as pointed out earlier, the communication between the membrane and the coupling factor is in fact by a stalk, the contribution of specific CF_1 subunits to this structure would suggest the design of specific experiments on the assembly of a proton-conducting channel into the protein itself.

The accessibility of CF_1 to antibodies against 4 of its subunits suggested to us that the protein could not possibly be buried within the membrane as some investigators have suggested. The burial notion was based mainly on negative experiments with thin sections and freeze-etch preparations which failed to reveal the

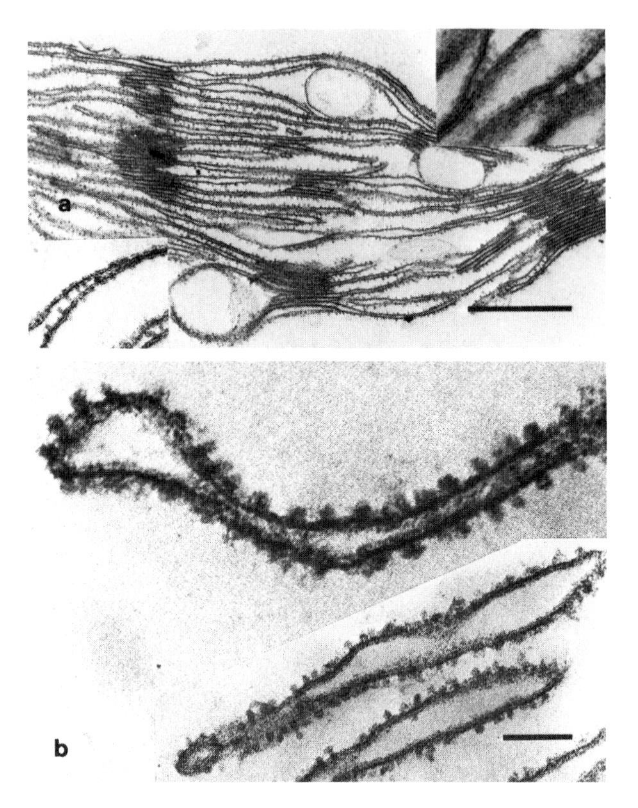

Fig. 4-4 Electron micrograph of chloroplasts stained with uranyl acetate in 100% ethanol. (Courtesy of Dr. E. Moudrianakis.) Experimental conditions were as described by Oleszko and Moudrianakis (1974). (a) Class II chloroplasts. Scale line, 0.5 μm. (b) Preparations of isolated disc. Scale line, 0.1 μm. Insets, stacked membranes are separating.

presence of the inner membrane spheres of F_1 (cf. Wrigglesworth *et al.*, 1970). However, as shown in Fig. 4-3 staining of fixed thin sections of mitochondria with methanolic osmium tetroxide revealed clearly the 85 Å spheres of F_1 (Telford and Racker, 1973) and similar pictures (Fig. 4-4) have been obtained with chloroplast by Oleszko and Moudrianakis (1974). Moreover, Garber and Steponkus (1974) have taken electron micrographs of freeze-etch preparations of chloroplasts which show characteristic inner membrane spheres. They have unambiguously demonstrated by resolution of the membrane with silicotungstate (S. Lien and Racker, 1971) and by reconstitution with CF_1 that the spheres were indeed CF_1 (Fig. 4-5).

Coupling Factor 2 (F_2, Factor B). This protein was shown to stimulate the P_i–ATP exchange in the presence of F_1 and OSCP in submitochondrial particles that have been treated with silicotungstate. An antibody prepared against factor B inhibited the exchange. Factor B and F_2 replaced each other in the respective assays and appeared to be identical (Racker *et al.*, 1970a). Curiously, in the assay of factor B in A-particles which depends on a stimulation of the energy-dependent reduction of DPN by succinate (Lam *et al.*, 1967), F_1 plus OSCP can substitute for factor B. Thus factor B does not function in this assay as a clearly defined and unique coupling factor. Nevertheless, the assay is most useful during the purification of F_2, provided F_1 and OSCP are not present in the sample to be analyzed.

Recently, a factor B or F_b has been isolated in homogenous form (Higashiyama *et al.*, 1975). However, the stimulation of DPN reduction by succinate by pure F_b is very minor (20–30%) compared to the four- to fivefold stimulation by partially purified F_2 or factor B. Moreover, no data are available on the effect of pure F_b on the P_i–ATP exchange.

The role of SH groups in F_2 or factor B has been emphasized previously (Racker and Horstman, 1967; Sanadi *et al.*, 1968), and the stability of the protein can be increased by dithiothreitol and 20% glycerol. There is no evidence for an F_2 or factor B requirement in the oligomycin-sensitive hydrolysis of ATP. Furthermore, no stimulation of either the $^{32}P_i$–ATP exchange or of oxidative phosphorylation by F_2 has been observed in reconstituted liposomes. It seems likely therefore that this factor renders the membrane less permeable to protons. This is in line with the observation that oligomycin at low concentrations substitutes for factor B in the assay with A-particles (Lam *et al.*, 1967). Nevertheless, it is possible that in addition to such a structural function

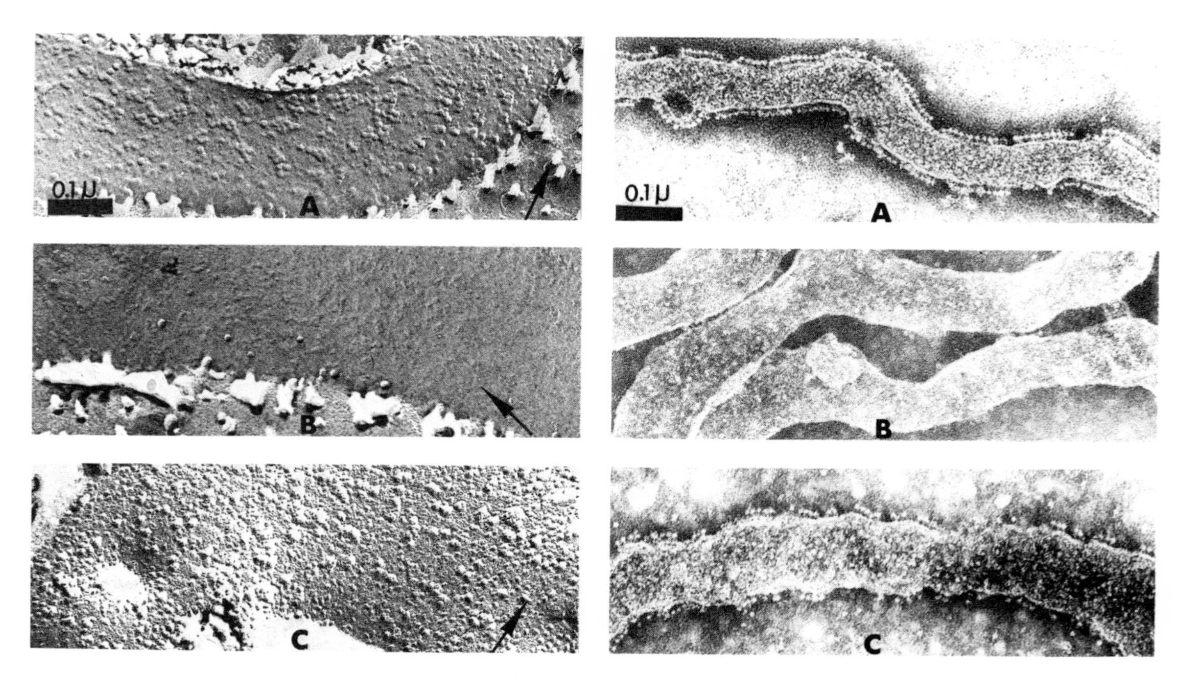

Fig. 4-5 Electron micrograph of freeze-etched silicotungstate-treated chloroplast particles before and after reconstitution with CF_1. Experimental conditions were as described by Garber and Steponkus (1974). Left, freeze-etched thylakoids; right, thylakoids negatively stained with phosphotungstate. (A) Untreated thylakoids; (B) silicotungstate-treated thylakoids; (C) treated thylakoids reconstituted with CF_1.

affecting proton permeability, F_2, like F_1, may also have a catalytic role in proton translocation. Precise answers to these questions must wait until pure F_2 which has all the properties of the crude coupling factor can be prepared. The observations of Higashiyama *et al.*, (1975) suggest that perhaps more than one protein is involved.

OSCP (Oligomycin Sensitivity Conferral Protein, F_4, Fc_1). When submitochondrial particles were exposed to sonic oscillation in the presence of 0.05 *M* NH_4OH, a coupling factor was released which was required for the oligomycin-sensitive hydrolysis of ATP (Kagawa and Racker, 1966a). First a crude preparation (F_4), later a purified protein called conferral factor (Fc or Fc_1) (Bulos and Racker, 1968a), was used for conferral of oligomycin sensitivity. MacLennan and Tzagoloff (1968) purified this factor from F_4 (or directly from mitochondria) and called it oligomycin sensitivity conferral protein (OSCP). It has a molecular weight of 18,000 and has a tendency to aggregate. It should be noted that OSCP or Fc_1 is only one of the proteins required for oligomycin sensitivity of the ATPase. It is not the component in mitochondria which reacts with oligomycin or DCCD (Knowles *et al.*, 1971). Its function in the ATPase complex will be discussed together with the role of the hydrophobic components. As mentioned earlier in most systems OSCP substitutes for F_3 and F_5. However, stimulatory effects by F_5 in the presence of OSCP have been recorded (Fessenden *et al.*, 1969; Fisher *et al.*, 1971b), suggesting either the presence of yet another coupling factor or of a factor counteracting an inhibitor. Nevertheless, until F_3 and F_5 are more clearly defined, we shall refrain from referring to them as distinct coupling factors.

Coupling Factor 6 (F_6, Fc_2). A heat-stable conferral factor (Fc_2) was shown to be required for the oligomycin-sensitive ATPase and the P_i–ATP exchange reaction (Racker *et al.*, 1969; Knowles *et al.*, 1971) in particles that had been exposed to silicotungstate or sodium thiocyanate. The factor appears to facilitate the binding of F_1 to the membrane (Knowles *et al.*,

1971). A highly purified preparation of the coupling factor (F_6), which was very sensitive to trypsin, contained several "isoproteins" which were active (Fessenden-Raden, 1972a). At least some of these "isoproteins" were caused by association of the protein with fatty acids which markedly affected their migration in acrylamide gels (Fessenden-Raden, 1972b). A homogenous preparation of F_6 was recently isolated from the oligomycin-sensitive ATPase (Kanner *et al.*, 1976). F_6 is the only heat-stable coupling factor of mitochondria and is therefore readily distinguished from the other components.

The Hydrophobic Components of the Oligomycin-Sensitive ATPase (Proteins and Phospholipids).

The reconstitution of the oligomycin-sensitive ATPase requires, in addition to coupling factors, a delipidated hydrophobic protein fraction, CF_0 (Kagawa and Racker, 1966a,b), and phospholipids. Reconstitution of a membrane with functional and morphological resemblance to the mitochondrial membrane was achieved by simply mixing the components (Kagawa and Racker, 1966c). I have reviewed these experiments in great detail earlier (Racker, 1970b) and shall only discuss some key features.

When F_1 was added to CF_0 the ATPase activity was completely inhibited but could be reactivated by addition of phospholipids. In contrast to oxidative phosphorylation, single pure phospholipids activated the masked ATPase complex (Bulos and Racker, 1968b). The inhibition of ATP hydrolysis by CF_0 did not involve the protein inhibitor of Pullman and Monroy (1963), but required OSCP which is present in the CF_0 preparation. When F_1 was added to a CF_0 preparation that was depleted with respect to OSCP, the ATPase activity became particulate (presumably bound to F_6 which is present in CF_0). This particulate ATPase activity did not require phospholipids and was insensitive to rutamycin or to DCCD. On addition of Fc_1 (or OSCP) ATP hydrolysis was inhibited but completely restored by phospholipids (Bulos and Racker, 1968b). This ATPase activity was now

TABLE 4-4 **Reconstitution of the Oligomycin-Sensitive ATPase**

Step I.	CF_0 (delipidated and OSCP-deficient) + F_1 = CF_0F_1 = Particulate oligomycin-insensitive ATPase
Step II.	CF_0F_1 + OSCP = CF_0F_1 OSCP = Particulate masked ATPase
Step III.	CF_0F_1 OSCP + phospholipids = Particulate oligomycin-sensitivy ATPase

sensitive to oligomycin or DCCD. These complex relationships are recapitulated in Table 4-4.

We have drawn the following two conclusions. Both F_6 and OSCP participate in the attachment of F_1 to the membrane (Knowles *et al.*, 1971). Phospholipids, particularly in the case of CF_1 (Livne and Racker, 1969), contribute to the binding also. However, at present there is no evidence whether OSCP or F_6 or both or neither represent the stalk seen in electron micrographs. The second conclusion is that neither hydrophobic protein nor OSCP alone inhibits the ATPase and that the interaction between these two components causes inhibition which is reversed by phospholipids. I therefore suggest that the inhibition of ATPase by energy transfer inhibitors such as oligomycin or DCCD is not a direct one but mediated by OSCP and the hydrophobic protein. An outline of such a mechanism is shown in Fig. 4-6. This formulation allows for an interpretation consistent with the chemiosmotic hypothesis with the stalk representing the pathway for proton translocation and OSCP blocking it in the absence of phospholipids.

Purification of biologically active CF_0 has progressed slowly (Chien, 1974; Serrano *et al.*, 1976). Virtually all respiratory contaminants have been removed. The major components seen in SDS-acrylamide gels is a band corresponding to a molecular weight of about 30,000 and bands in the region of 10,000 to 13,000. While we have exclusively relied on biological assays for the isolation of active components, other investigators have used the binding of radioactive DCCD as an assay (Cattell *et al.*, 1970; Stekhoven *et al.*, 1972). Consistent with the finding that neither

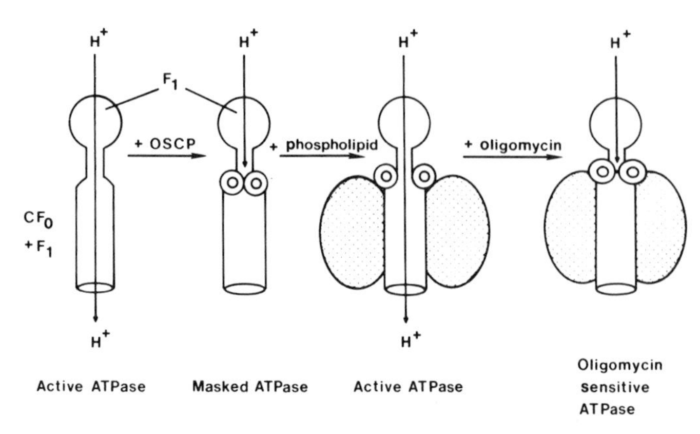

Fig. 4-6 Schematic representation of different stages in the reconstitution of oligomycin-sensitive ATPase.

F_1, OSCP, nor F_6 (Fc_2) are the sites at which energy transfer inhibitors interact (Knowles *et al.,* 1971), the studies with radioactive DCCD point to a small proteolipid of about 12,000 molecular weight as the reactive species. Our attempts to show that this proteolipid is active in either $^{32}P_i$–ATP exchange or oxidative phosphorylation have thus far failed. It is therefore possible that the long treatment with chloroform–methanol has split the proteolipid off the larger (30,000 MW) complex and perhaps both components are required for the biological activity. Recently, a water-soluble preparation of mitochondrial proteolipid was shown to act as a proton ionophore (Racker, 1975b), but thus far oligomycin sensitivity has not been conferred to this system.

The ATP-Driven Proton Pump

According to Mitchell (1966) an ATP-driven proton pump functions as the coupling device. Proton movements on addition of ATP to mitochondria or to submitochondrial particles have been recorded (Mitchell and Moyle, 1965; Chance and Mela, 1967). Quantitative measurements of this process in mitochondria are complicated by movements of other ions associated with proton movements and by the fact that ATP has to be transported to the matrix side of the membrane where F_1 is located. In submitochondrial particles which have F_1 on the outside, measurements of proton movements in the presence of ATP are complicated by the fact that hydrolysis of ATP by F_1 is associated with large changes in pH. In elegant studies Thayer and Hinkle (1973a) eliminated this difficulty by measuring proton translocation at pH 6.2 at which ATP hydrolysis takes place without pH changes. Under these conditions close to 2 H^+ were translocated for each ATP hydrolyzed. Both proton movements and ATP hydrolysis were inhibited by oligomycin. Recent studies of ATP-driven proton movements in chloroplasts have yielded H^+/ATP ratios approaching 4 (Carmeli *et al.*, 1975).

The proton pump of mitochondria and chloroplasts is capable of ATP formation. Jagendorf and Uribe (1966) have demonstrated ATP formation in chloroplasts during an acid to base transition. McCarty and Racker (1966) have shown coupling factor-dependent ATP formation during acid to base transition in subchloroplast particles. Thayer and Hinkle (1973b, 1975) observed pH transition-dependent ATP formation in submitochondrial particles in the presence of a membrane potential. In careful kinetic experiments they have shown that the rate of ATP formation by a pH gradient is actually faster than the rate during oxidative phosphorylation. This established the kinetic competence of the proton pump as a legitimate candidate for the coupling device. In chloroplasts Schuldiner *et al.* (1972), Avron *et al.* (1973), and Portis and McCarty (1974) have collected impressive evidence on the relationship between the proton gradient and ATP formation.

It is well established, therefore, that the proton pump of mitochondria and chloroplasts can generate ATP and ADP and P_i when provided with a proton flux, and that the rate of this process is compatible with the requirements of a participation in oxidative phosphorylation. I shall return in Lecture 6 to the proton pump of mitochondria and its components when I describe its reconstitution.

Lecture 5

The Oxidation Chain and the Topography of the Inner Mitochondrial Membrane

Was it bullets or a wind
or a rip-cord fouled on chance
Artifacts the natives find
Decorate them when they dance.

John Ciardi, *Elegy Just in Case*

The slightest blow will break a dish that's cracked.

Ovid, *Tristia*

The main feature of the oxidation process in mitochondria and chloroplasts is that it takes place in multiple steps. In mitochondria the reactions proceed down the thermodynamic ladder as shown in Fig. 5-1 from DPNH (−0.32 volt) to oxygen (+0.82 volt) catalyzed by a multienzyme system often referred to as the respiratory chain. Since chloroplasts do not respire (cf. Fig. 2-2), the name oxidation chain is preferable. If oxidation would take place without any provisions of energy conservation and transduction, all the potential energy of oxidation would be lost in heat. In mitochondria and in chloroplasts, electron transport is held in check by a mechanism called respiratory or oxidation control. Again, the latter name is better.

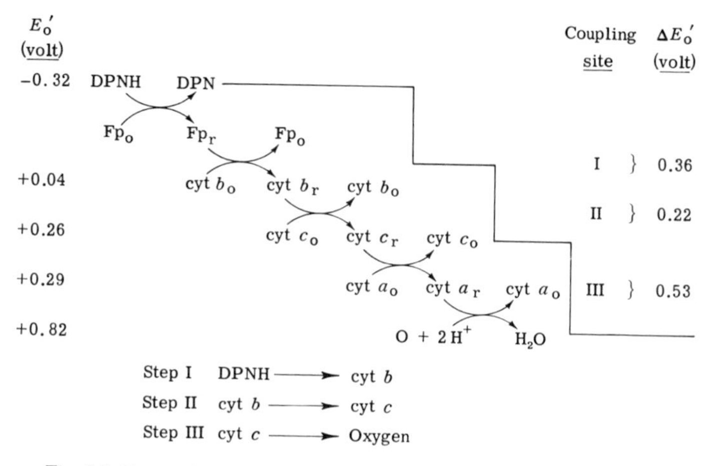

Fig. 5-1 Energetics of the mitochondrial oxidation chain. Subscript "o" means oxidized; subscript "r" means reduced.

Analysis of the Oxidation Chain

The elucidation of the oxidation chain in mitochondria can be attributed primarily to the efforts of David Keilin who discovered between 1925 and 1930 the cytochrome-abc of the oxidation chain with the aid of a hand spectroscope (Keilin, 1925). Actually MacMunn had discovered cytochromes 40 years earlier, but was discredited by a powerful scientist named Hoppe-Seyler. MacMunn (1914) wrote: "A good deal of discussion has taken place over this pigment and the name of Hoppe-Seyler has prevented the acceptance of the writer's views. The chemical position is undoubtedly weak but in time this pigment will find its way into the textbooks." It has, indeed.

> *Lesson 12:* If you are a biologist be sure your chemistry is strong or the chemists won't believe you. If your chemistry is strong they will believe you even if you are wrong.

Analysis of spectral changes during oxidoreduction have come a long way since the discovery of cytochromes. Brilliantly designed dual wavelength spectrophotometers (Chance and Williams, 1956), flashes generated by liquid dye lasers, and recordings on oscilloscopes with data channeled through computers (Chance, 1972a) have allowed kinetic analyses of individual steps. Inhibitors have served as important aids in these studies as illustrated by the example of antimycin. When this inhibitor is added to mitochondria, some of the carriers become more reduced, others more oxidized. The "crossover point" is between cytochrome *b* and cytochrome *c*. Chance and Williams (1956) have analyzed in elegant studies the control of respiration by determining the crossover points of oxidative phosphorylation as revealed by spectral changes on addition of ADP which releases the oxidation control. By using appropriate inhibitors three crossover points corresponding to the three sites of oxidative phosphorylation were described. Because of overlapping regions of absorption spectra, e.g., between flavoproteins and nonheme iron protein, the method is not without its pitfalls and ambiguities. Moreover, the use of inhibitors is often treacherous and almost invariably complications arise when their precise site of action is not known, as in the case of rotenone which will be discussed later.

> *Lesson 13:* If you accept the statement that only uninhibited investigators use inhibitors, you will soon find out what kind of people work in the field of oxidative phosphorylation.

Isolation of Complexes and Individual Catalysts

With substrates such as succinate and ascorbate which enter the oxidation chain at lower checkpoints, segments of the multienzyme system can be analyzed in intact mitochondria and submitochondrial particles. Hatefi *et al.* (1962) made important contributions by fractionating the oxidation chain into individual complexes which catalyze discrete segments of the oxidation chain. These complexes, which I shall discuss in greater detail, greatly simplified the approach to the identification of the catalytic components and the determination of the sequences in the electron transfer. They also serve as valuable starting material for the isolation of active and pure catalytic components.

Considerable efforts have been expended in the past to isolate the individual catalysts from mitochondria or from the complexes. Cytochromes and flavoproteins have been obtained that were spectrally pure, i.e., they had a single chromophore attached to the protein and they were homogenous by other criteria. For the isolation of these proteins it was simple and convenient to use spectral changes that take place in the presence of artificial reductants or oxidants as an assay during purification. If there is a simple assay available for anything, you can be sure somebody will purify it.

> *Lesson 14:* Many of us have been led astray by experiments that were "so easily done." Where there is an easy assay, there is a way, there is a will. If we could only look ahead and see how long the easy road is and where it will lead us!

It was rather unfortunate that spectral changes of the chromophores of respiratory catalysts were used as the only assay during purification. Cytochromes and flavoproteins have been isolated and numerous papers published on kinetics, physical properties, and mechanism of action. Some of this work has not only been of little use, but is actually misleading because the catalysts had been severely altered during the isolation procedure. An instructive example is succinate dehydrogenase. The early preparations of this enzyme were chemically degraded and incapable of interaction with the membrane, yet their catalytic activity as a dehydrogenase with succinate as electron donor and artificial electron acceptors was unimpaired. King (1963) showed that the presence of succinate during the purification of the protein is a decisive factor in the isolation of a reconstitutively active enzyme. It seems imperative that the modern membrane protein chemists relate protein structure not only to the catalytic and allosteric sites of the enzyme, but also to the allotopic site which is responsible for its integration within the membrane and reactivity with its natural neighbors.

Lesson 15: Hair will grow on a dead body and spectral shifts take place on dead proteins. When you purify an enzyme make sure it stays alive and active within its biological community.

Since throughout these discussions I shall emphasize the functional aspects of the oxidation chain which include coupled phosphorylation, it seems appropriate to divide our discussion into three segments each responsible for the generation of one ATP from ADP and P_i for each electron pair that passes through.

The Three Segments of the Oxidation Chain

The First Segment (DPNH to Coenzyme Q)

A very active complex I was isolated by Hatefi *et al.* (1962). It catalyzes the rotenone-sensitive reduction of Q_{10} or Q_1 by DPNH. Hatefi and Stempel (1967) resolved two protein components from complex I, a DPNH dehydrogenase flavoprotein and a nonheme iron protein which reacted sequentially, but in view of their sluggish interactions the question has been raised whether they represent physiological events (Singer and Gutman, 1970). Two types of DPNH dehydrogenase preparations have been isolated from mitochondria (Salach *et al.,* 1967). The first type reacts with ferricyanide as an acceptor. The second type, which can be obtained by exposing the first type, e.g., to heat or acidic ethanol, reacts with many acceptors including coenzyme Q analogues and cytochrome *c*. However, in contrast to the reduction of these natural acceptors within intact mitochondria or submitochondrial particles, the reactions catalyzed by the isolated dehydrogenase are insensitive to rotenone. There can be little doubt that this more reactive species of the dehydrogenase which has been prepared by a variety of relatively harsh methods and which has been studied in numerous laboratories, represents a severely altered enzyme. It can be fooled into reacting with natural electron acceptors which normally do not interact directly with the protein. Unfortunately even the more gently

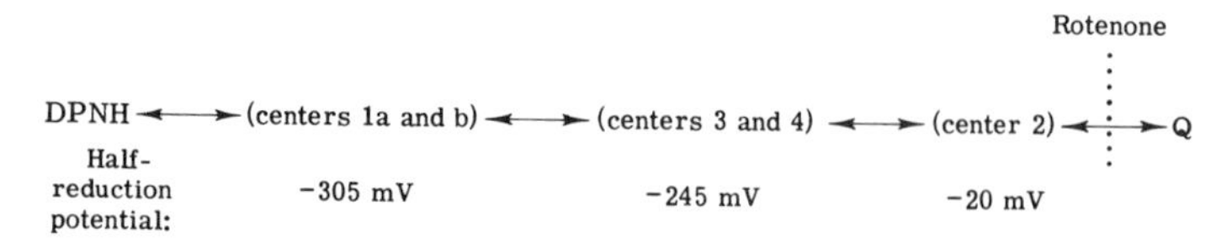

Fig. 5-2 The first segment of the mitochondrial oxidation chain and the centers of DPNH dehydrogenase.

prepared DPNH dehydrogenase that reacts with ferricyanide as acceptor is inactive in reconstitution of oxidative phosphorylation, whereas a reconstitutively active DPNH dehydrogenase has been prepared by delipidation of complex I (Ragan and Racker, 1973b). This dehydrogenase catalyzed a rotenone-insensitive reduction of ferricyanide but did not reduce coenzyme Q_1. An active complex was reconstituted by mixing this preparation with pure phospholipids in the presence of cholate followed by dialysis. The vesicles thus formed catalyzed not only the rotenone-sensitive reduction of coenzyme Q by DPNH, but when reconstituted in the presence of the coupling device, catalyzed oxidative phosphorylation.

Experiments with various yeast strains that were grown in the presence of limiting amounts of iron and sulfur implicated nonheme iron and labile sulfide groups in electron transport and energy conservation in the first oxidation segment (cf. Ragan and Garland, 1971; Ohnishi, 1973). Sophisticated and penetrating studies on the electron paramagnetic resonance spectra at liquid helium temperature have revealed multiple nonheme iron centers in the first oxidation segment of submitochondrial particles from yeast (Ohnishi *et al.*, 1970) and from bovine heart (Albracht and Slater, 1971; Orme-Johnson *et al.*, 1971a). A thermodynamic profile for these centers essentially as proposed by Ohnishi *et al.* (1972) is shown in Fig. 5-2. This scheme is consistent with the interesting experiments of Gutman *et al.* (1971, 1972) who analyzed changes in absorbance at 470 minus 500 nm in submitochondrial particles. A rapid initial bleaching on addition of DPNH was followed by return of color when DPNH was completely oxidized. Center 2, which was identified as the site of these changes in the nonheme iron chromophores, was rapidly oxidized on addition of ATP. In the presence of rotenone or piericidin, reoxidation was inhibited. Uncouplers prevented the ATP-dependent reoxidation of the chromophore. Based on these observations and on previous studies by Hinkle *et al.* (1967b), the assignment for the site of rotenone inhibition is shown in Fig. 5-2 on the oxygen side of the nonheme iron of site 1. Some of the controversy regarding the location of the rotenone inhibition site was based mainly on rotenone-sensitive absorbancy changes which were ascribed to nonheme iron (Hatefi, 1968). However, we have shown that the observed changes were caused by reduction of cytochrome *c* and were abolished on addition of excess cytochrome oxidase (Ragan and Racker, 1973b).

Attempts have been made to identify a specific energy transducing component at site 1. From low temperature EPR measurements of the phosphate potential dependence of the half-reduction potential, a tentative assignment to the iron–sulfur center 1a has been made (Ohnishi, 1973). Later experiments implicated also center 2 which exhibited a more positive half-reduction potential on addition of ATP. Such designation of specific-energy transducing components of the oxidation chain have a meaning only in the context of the chemical hypothesis. As I have pointed out in the discussion of the high-energy intermediates of oxidative phosphorylation, the energy-dependent shifts in the half-reduction potential of the electron transport carriers are subject to a different interpretation as was first pointed out by Hinkle and Mitchell (1970).

The Second Segment (QH_2 to Cytochrome c_1)

Complex III (Hatefi *et al.*, 1962) catalyzes the reduction of cytochrome *c* by reduced Q. It contains cytochrome *b*, cytochrome c_1, a nonheme iron protein, and an oxidation factor required for the reduction of cytochrome c_1 by cytochrome *b* which was described by Nishibayashi *et al.* (1972). This factor which has been purified is a rather labile protein and does not have the EPR signal at $g = 1.90$ (B. Trumpower, unpublished observations) characteristic of the reduced nonheme iron protein described by Rieske *et al.* (1964). What could be the role of this factor in respiration? According to the chemiosmotic formulation, each oxidation segment or loop must have a proton carrier. Mitchell (1966) has proposed Q_{10} as proton carrier in the second oxidation segment, but most of the experimental evidence points to Q_{10} participation in the chain before cytochrome *b*. It is conceivable that the new oxidation factor fulfills the role of a hydrogen carrier in the segment between cytochrome *b* and c_1. An alternative role for this factor will be proposed below.

There are multiple forms of cytochrome *b* in the second segment (Slater *et al.*, 1970; Erecinska *et al.*, 1972). Three distinct absorption maxima at 562, 566, and 558 (558.5, 562, and 554.5 at 77°K) can be discerned (cf. Berden, 1972). Cytochrome *b*-566 has also been referred to by Chance and his collaborators as cytochrome b_T (energy transducing) and cytochrome *b*-562 as cytochrome b_K. It was proposed (Erecinska *et al.*, 1972) that during electron transport cytochrome b_T becomes high potential and can be stabilized by antimycin. It also appears that oxidation of cytochrome c_1 is required for the reduction of cytochrome b_T (Rieske, 1971; Chance, 1972b). This is an interesting observation perhaps related to the phenomenon of respiratory control. EPR signals at $g = 3.44$ and 3.78 corresponding to the cytochrome *b* species 562 and 566 have been observed in complex II (Orme-Johnson *et al.*, 1971b).

In a lucid review Wikstrom (1973) discussed the confusion in this area of research caused partly by methodological variations in absorption measurements, partly by different nomenclatures, and partly by misinterpretation of data. He recommends that we should refer to cytochrome *b*-562 and cytochrome *b*-566 (which includes an absorption shoulder at 558 nm) as the two major mitochondrial components and avoid designations such as b', b_c', b_T, and b_K, which have been used by various investigators. I concur with his recommendation particularly since the designation b_T implies the existence of a specific energy transducing component of the oxidation chain, a concept that is not shared universally. It was, moreover, pointed out by Slater (1973) that

ATP has an effect on the midpoint potential of *b*-562 as well as on *b*-566. He also recommends that the designation b_K and b_T be abandoned.

There is no consensus on the specific functions of the different cytochrome *b* species. Wikstrom (1973) suggests an asymmetric distribution with one cytochrome *b* complex on one side of the membrane and another on the other side. This would be analogous to the distribution of cytochrome *a* and a_3. This formulation is similar to that of Mitchell (1972) and satisfies the chemiosmotic hypothesis which requires transmembranous electron flow. However, from kinetic analyses cytochrome *b*-566 and *b*-562 are assigned by Wikstrom to both sides of the membrane which does not help to explain the needed topographical asymmetry and functional dichotomy. Moreover, in his scheme there is no role for the oxidation factor which participates between *b* and c_1.

There is a great need for further studies on the topography of the cytochrome *b* species. If as suggested by Wikstrom (1973) there is some preponderance of *b*-566 on the C-side of the mitochondrial membrane and of *b*-562 on the M-side, this might be more firmly established by studies with the aid of chemical derivatization as performed on the third segment of the oxidation chain which will be discussed later. One would hope that more significant data could be obtained with isolated complexes and reconstituted systems, particularly since complex III was recently successfully incorporated into vesicles which exhibit respiratory control (Hinkle and Leung, 1974) or catalyze oxidative phosphorylation (Racker *et al.*, 1975b). It is therefore significant that complex III (Hatefi *et al.*, 1962) contains both cytochrome *b*-562 and *b*-566. The presence of a high potential *b*-562 in complex III (Berden, 1972) is of questionable significance since this component cannot be detected in some intact mitochondria.

Reduced Q_{10} which is the substrate for the second segment of the oxidation chain can be generated either from DPNH by complex I or from succinate by complex II. Complex II contains succinate dehydrogenase and a cytochrome *b* with an absorbance maximum at 556 nm at 77°K (Davis and Hatefi, 1971). The significance of cytochrome *b* in complex II has been questioned

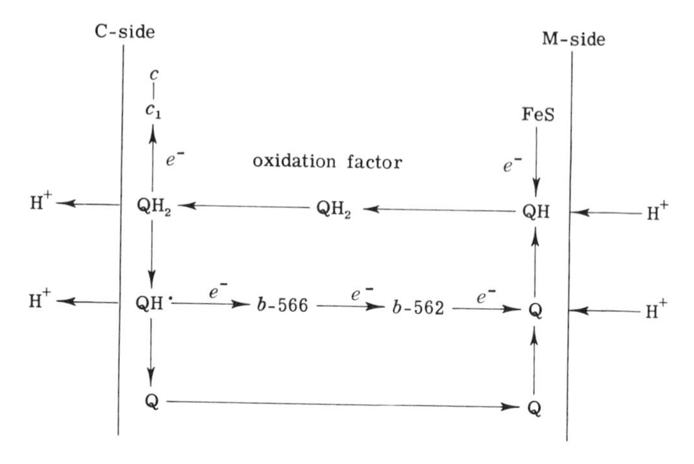

Fig. 5-3 The Q cycle in the second segment of the mitochondrial oxidation chain [loop 2 + 3 of Mitchell (1966)].

(Berden, 1972), and contamination by a modified cytochrome *b* was suggested. On the other hand, the resolution and reconstitution of complex II (Bruni and Racker, 1968) yielded suggestive evidence that, in the pathway from succinate to Q_{10}, cytochrome *b* may play a structural role which can be satisfied by cytochrome *b* that had been exposed to sodium dodecyl sulfate and does not seem to turn over during Q_{10} reduction.

A rather ingenious scheme recently has been proposed by Mitchell (1975) which explains many of the perplexing features of this oxidation segment. As shown in Fig. 5-3 the key feature of this scheme is the participation of a Q_{10} radical in an oxidoreduction cycle and a cycling of electrons resulting in an H^+/e ratio of 2 in this segment. This scheme accounts, as pointed out by Mitchell, for the multiple cytochrome *b* requirements, for the known interdependence between cytochrome *b* reduction and cytochrome c_1 oxidation and for some of the curious observations made with antimycin. It also points to the need of an additional catalyst, namely, for the transfer of electrons from QH_2 to cytochrome c_1 which would be a logical function for the oxidation factor described earlier. The scheme shown in Fig. 5-3 is essentially that described in the publication of Mitchell (1975) with some modifications. The entry and release of protons is clearly assigned to Q or its derivatives. The original scheme is ambiguous on this point and leaves the impression that the entry of H^+ on the M-side is at cytochrome b_K (b-562) which would require that b_K donates hydrogens to Q while in the scheme shown in Fig. 5-3 the cytochrome functions only as an electron donor. I have also substituted b-562 and b-566 for b_K and b_T. But the major change from the original scheme was suggested by P. Mitchell (personal communication). QH_2 instead of $QH^•$ donates electrons to c_1, and $QH^•$ rather than QH_2 donates electrons to cytochrome b. We have performed experiments which suggest that the oxidation factor in fact catalyzes the reduction of cytochrome c_1 by reduced Q.

The Third Segment (Ferrocytochrome c to Oxygen)

Cytochrome oxidase is a multisubunit enzyme which catalyzes the oxidation of reduced cytochrome c by molecular oxygen. The enzyme contains two spectrally distinguishable cytochromes (a and a_3). Yeast cytochrome oxidase consists of seven subunits (Schatz and Mason, 1974). The three larger subunits are of mitochondrial origin; the smaller subunits are synthesized by cytoplasmic ribosomes and incorporated into the mitochondrial membrane. It is of great interest that antibodies against small as well as large subunits inhibit cytochrome oxidase activity (Poyton and Schatz, 1974). This observation confirms and extends the conclusion drawn from experiments with mutants (Schatz and Mason, 1974). Thus far, the roles of the individual subunits have not been established. One of the approaches to this problem is to study the reconstitution of cytochrome oxidase into phospholipid vesicles which will be discussed in Lecture 6. Thus far, only cytochrome oxidase preparations that were prepared with cholate (and not with deoxycholate) have been suitable for reconstitution. Enzyme preparations that were exposed to deoxycholate during purification are poorly suited for reconstitution. It is unfortunate that so many interesting physical studies (Vanderkooi *et al.*, 1972) have been performed with reconstitutively inactive cytochrome oxidase preparations. A great deal has been written about the mechanism of action of cytochrome oxidase, but little of it can be related at present to problems of energy generation in the third segment of the oxidation chain. The significance of energy-dependent absorption changes in this region has been discussed in Lecture 3.

The Topography of the Oxidation Chain

The Proteins

With a few exceptions (Green and Ji, 1972) the majority of investigators in the field agrees that the oxidation chain is organized in an asymmetric manner (Racker, 1970a). The evidence for asymmetry derives from work done in several laboratories. Diverse methods have been used to assess the localization of the membrane components. Among the most informative approaches are the use of (a) specific antibodies against individual components, (b) impermeant chemicals which serve as modifiers of the surface components of the membrane, or (c) impermeant artificial electron acceptors or donors. A few examples should suffice to illustrate the types of experiments which have been conducted with these tools.

The study of the topography of the inner mitochondrial membrane has been possible because of the reorganization of the membrane which takes place during sonication. The mitochondrial cristae break off in such a manner that the resulting submitochondrial particles are inside out (Lee and Ernster, 1966; Mitchell, 1966). Thus, the mitochondrial ATPase (F_1), which in mitochondria faces the matrix, faces the medium in submitochondrial particles. In fact, it can be stated that in such preparations all submitochondrial particles that are functionally active in oxidative phosphorylation are inside out, because phosphorylation can be completely inhibited by an antibody against F_1 (Fessenden and Racker, 1966). Although during sonication of mitochondria there is some dislocation of cytochrome oxidase, which

TABLE 5-1 **Effect of Antibodies against Mitochondrial Membrane Components**

Antibody	Test	Mitochondria	SMP
Anti-F_1	P:O	No inhibition	Inhibition
Anti-cytochrome *c*	Succinate oxidation	Inhibition	No inhibition
Anti-cytochrome c_1	Succinate oxidation	Inhibition	No inhibition
Anti-cytochrome oxidase	Reduced cytochrome *c* oxidation	Inhibition	Inhibition

I shall discuss shortly, there is no evidence that the dislocated protein is transmembranous as would be expected if some submitochondrial particles were oriented right side out as in mitochondria. Furthermore, we have failed to detect any ATPase inside of submitochondrial particles.

Antibodies against respiratory components (Racker *et al.*, 1970b) inhibit the functional activities of mitochondria and submitochondrial particles in a characteristic manner as shown in Table 5-1. These data confirm the conclusion from various experiments (cf. Mitchell, 1966) that cytochrome *c* is located on the outer side (C-side) of the mitochondria. They also show that cytochrome c_1 is accessible to antibodies added to the C-side of the membrane. Results with antibodies directed against cytochrome oxidase suggested that this component is accessible from both sides of the membrane. The interpretation of these findings is, however, complicated by the fact that ferrocytochrome *c* is oxidized by submitochondrial particles as well as by mitochondria. This is caused by some displacement of cytochrome oxidase during sonication so that cytochrome *a* becomes available on both sides of the membrane. The question then arises whether the cytochrome oxidase is actually transmembranous in both directions or only in one. Experiments conducted by Dr. C. Carmeli have shown that, in contrast to reconstituted site 3 vesicles (Racker and Kandrach, 1973) which contain cytochrome oxidase oriented in both directions, in sonicated submitochondrial particles, transmembranous cytochrome oxidase at least functionally is unidirectional (inside-out topography). The displaced cyto-

TABLE 5-2 **Incorporation of Diazobenzene Sulfonate into Surface Components of the Mitochondrial Membrane**[a,b]

	$[^{35}S]$ DABS (cpm)		
Component	Isolated from mitochondria	Isolated from SMP	Soluble protein
F_1	80	1514	1450
Cytochrome *c*	758	100	783
Cytochrome oxidase	265	197	1580

[a]Experimental conditions were as described by Schneider *et al.* (1972).

[b] $N{\equiv}\overset{+}{N}-C_6H_4-SO_3^-$ **interacts with NH_2, SH, OH, etc.**

chrome oxidase is incapable of catalyzing either oxidative phosphorylation or uncoupler-sensitive pH changes during ascorbate oxidation. The displaced cytochrome oxidase may have arisen by some solubilization of the enzyme and readsorption to the submitochondrial particles. This proved to be a serious complication in the analysis of the subunit topography of cytochrome oxidase. As will be described later this made it necessary to prepare submitochondrial particles that have little or no displaced cytochrome oxidase.

A second approach to the topography is the chemical derivatization of the membrane by nonspecific but impermeant reactants such as ^{35}S-labeled diazobenzene sulfonate (Schneider *et al.*, 1972). These studies confirmed the conclusions drawn from experiments with antibodies, and in addition gave some quantitative information. As shown in Table 5-2 the surface components of the membrane were rendered radioactive to an extent comparable to that of the soluble protein. The availability of F_1 to the reagent confirms the conclusion that this protein cannot be deeply buried inside the membrane. On the other hand, cytochrome oxidase isolated after exposure of the membrane to diazobenzene sulfonate was much less radioactive than the exposed solubilized enzyme, suggesting that a large portion of cytochrome oxidase is indeed buried within the membrane. Based on these experiments and those described below, a transmembranous location of the enzyme was proposed.

Cross-linking of mitochondria with polylysine which reacts with cytochrome *a* (Schneider *et al.*, 1972) completely eliminated the reduction of either cytochrome *a* or a_3 by succinate while cytochrome *b* became reduced, thus permitting the conclusion that all cytochrome *a* is on the C-side and that no succinate-reducible cytochrome *a* is associated with the M-side of the membrane. The asymmetry of the organization of cytochrome oxidase across the membrane became strikingly apparent from studies of the location of its subunits discussed below.

A third approach to the topography of the catalysts of the oxidation chain is to expose mitochondria and submitochondrial particles to an electron acceptor or donor which cannot penetrate from one side to the other, e.g., ferricyanide (Tyler, 1970; Klingenberg, 1971). Depending on the assumptions made, this approach has led to conflicting conclusions about which components of the electron transport chain interact directly with ferricyanide. Thus, Tyler (1970) concluded that cytochrome *c* and c_1 are located on the C-side consistent with the results obtained by the immunological method, but he placed cytochrome *a* on the M-side of the membrane which is contrary to our findings. On the other hand, Klingenberg (1971, Fig. 8) placed cytochrome *a* on the C-side of the membrane, but cytochrome c_1 was not believed to be accessible from the C-side. In view of these divergent interpretations of the ferricyanide experiments, we have relied more on the approach with antibodies and chemical derivatization.

A recent analysis of the subunit topography of cytochrome oxidase with radioactive diazobenzene sulfonate gave rather convincing support to the location of cytochrome *a* and a_3 suggested above (Eytan *et al.,* 1975a). As shown in Fig. 5-4, subunits II, V, and VI are located on the C-side of the membrane accessible to DABS in mitochondria (and reconstituted vesicles). Subunit III is located at the M-side of the membrane. To obtain interpretable

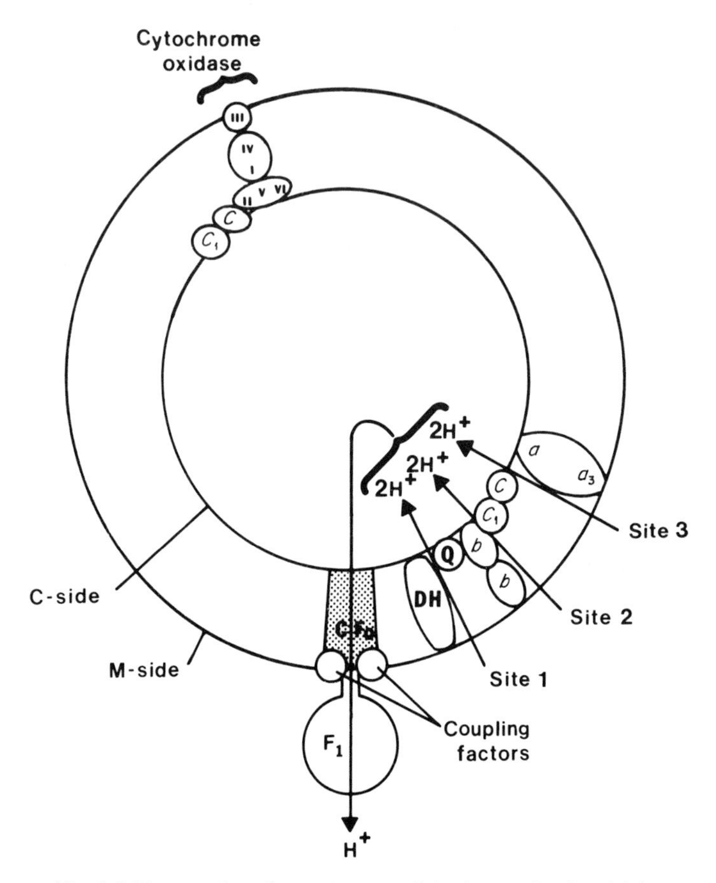

Fig. 5-4 Topography of components of the inner mitochondrial membrane.

data on the last point it was necessary to isolate submitochondrial particles that contained little or no displaced cytochrome oxidase. Such particles were obtained (though in low yield) by removing particles which contained cytochrome *a* on the surface by precipitation with antibodies. It appears that antibodies formed against cytochrome oxidase react preferentially with submitochondrial particles which contain displaced cytochrome oxidase.

A summary scheme of the topography of the inner mitochondrial membrane is shown in Fig. 5-4. F_1 and succinate dehydrogenase are located on the M-side of the membrane facing the matrix together with the other coupling factors. Cytochrome *c* and cytochrome c_1 are located on the C-side of the membrane facing the outer mitochondrial membrane. Cytochrome oxidase is transmembranous with cytochrome *a* on the C-side and cytochrome a_3 on the M-side.

The Phospholipids

Experiments on the effect of phospholipases on mitochondria and submitochondrial particles (Burstein *et al.*, 1971a,b) revealed that a relatively large portion of the phospholipids can be digested from either side of the membrane without serious loss of phosphorylation capacity. It seems that some of the external phospholipids are associated with secondary processes such as calcium transport, which was seriously impaired after digestion. The presence of accessory phospholipids on the surface complicates an analysis of the topography and the role of specific phospholipids in the operation of the oxidation chain and of the coupling device. This is unfortunate since the asymmetric organization of the membrane may be dependent on a phospholipid asymmetry. This notion is supported by the multiple phospholipid requirement for the reconstitution of oxidative phosphorylation (Racker and Kandrach, 1973) as well as of cytochrome oxidase vesicles with respiratory control (Racker, 1972b). The coupling device also requires both phosphatidylcholine and phosphatidylethanolamine (Kagawa *et al.*, 1973a). It was shown in experiments with the phosphatidylcholine exchange (or transfer) enzyme (Kagawa *et al.*, 1973b) that phosphatidylcholine can be added to the surface of preexisting phosphatidylethanolamine vesicles, thereby conferring to them the ability to catalyze P_i–ATP exchange. In view of the low incidence of the flip-flop between the phospholipids of the two phospholipid layers (Kornberg and McConnell, 1971; Johnson *et al.*, 1975), it appears that the coupling device can operate with little or no phosphatidylcholine in its inner membrane layer. It will be of interest to perform the reverse experiments when a purified phosphatidylethanolamine transfer enzyme becomes available. The phosphatidylcholine exchange enzyme is being used by Johnson *et al.* (1975) for a quantitative analysis of the surface distribution of phosphatidylcholine and also for a reevaluation of the flip-flop rate. Other methods, particularly involving NMR, are being developed at present for the analysis of the phospholipid topography.

An interesting study on the asymmetry of reconstituted cytochrome oxidase vesicles with respiratory control is now being conducted by R. Carroll and P. DiCorleto at Cornell University. They find a pronounced asymmetry in the reconstituted liposomes with phosphatidylcholine predominating on the C-side of the vesicles.

Finally, there will be need in the future to develop methods for the evaluation of protein–protein and protein–phospholipid interaction within the membrane. Only then will we be able to construct a three-dimensional structure of the membrane to aid us in the understanding of its architecture.

Lecture 6

Resolution and Reconstitution of Oxidative Phosphorylation

> If you have an important point to make, don't try to be subtle or clever. Use a pile driver. Hit the point once. Then come back and hit it again. Then hit it a third time—a tremendous whack.
>
> **Winston Churchill**

> There is only one answer to these objections: they are unfounded.
>
> **Paul Hindemith,**
> *Elementary Training for Musicians*

General Comments

A search of the literature on the resolution of membrane components reveals that investigators have used two basically different approaches. The first, more commonly used, is to select a membrane component which has a specific property, e.g., an enzyme activity, a characteristic spectral property, or exhibits a unique interaction with an inhibitor which provides an assay during purification. In the case of the mitochondrial membrane this approach has been used to isolate individual members of the respiratory chain, the mitochondrial ATPase, and proteins that specifically interact with DCCD or with atractyloside.

The advantage of this approach is simplicity and usually accuracy. The disadvantage is that one can never be certain that the purified component has retained its capacity to interact with neighbors within the membrane community, i.e., that it is reconstitutively active.

As I mentioned in my last lecture small variations in the procedures for the isolation of succinate dehydrogenase or cytochrome oxidase determine whether or not the enzyme is biologically competent. Another example is the proteolipid of molecular weight of about 12,000 which interacts with DCCD (Cattell *et al.,* 1970). Thus far, attempts to show that the proteolipid prepared by the described procedure is reconstitutively active have failed. On the other hand, a biologically active complex has been isolated (Chien, 1974; Serrano *et al.,* 1976) which contains multiple components, one of which migrates in SDS-acrylamide gels corresponding to a molecular weight of about 30,000, the others corresponding to molecular weights of 10,000 or even lower.

The second approach to the isolation of membrane components relies on an assay of biological functions within a more complex multienzyme system. There are various degrees of sophistication in the concept of reconstitution. The complexes of respiration (Hatefi *et al.,* 1962) catalyze, when properly assembled, the entire chain of oxidation but do not catalyze oxidative phosphorylation. The oligomycin-sensitive ATPase can be assembled to catalyze proton translocation (Kagawa *et al.,* 1973a) but does not catalyze respiration. The advantage of reconstituting such small segments of a pathway is not only that more specific information about the required ingredients can be obtained but also the risk of running into artifacts is not as great as with the first approach.

In the final analysis the crucial test for the biological activity of a membrane component is its capability to function within the overall physiological process, in our case, oxidative phosphorylation. The disadvantage of this most demanding approach is that the assay is usually quite complex and often less accurate. It requires membrane preparations that are either deficient in a single component or can be supplemented with all the other needed constituents in highly purified form.

It is therefore not surprising that, in the early stages of the resolution of oxidative phosphorylation, conflicting results were obtained not only in different laboratories but even within the same laboratory when different preparations of particles and factors were recombined under varying conditions. In spite of this serious disadvantage the final analysis must in all cases be carried out in the complete functional system. However, we have often resorted, as mentioned in the first lecture, to using an abbreviated assay such as the oligomycin-sensitive ATPase or the energy-linked reduction of DPN by succinate as a routine assay during purification of a component, checking from time to time its capability to function in oxidative phosphorylation.

Reconstitutions of Mitochondrial and Chloroplast Membrane Functions

Apoparticles and Coupling Factors

The resolution of various components of the coupling device of oxidative phosphorylation was first achieved by a gentle approach which we refer to as resolution from without. Submitochondrial particles were exposed to mild physical or chemical treatment that resulted in the dissociation of one or more surface components (Pullman *et al.,* 1960). The depleted membranes or apoparticles were now incapable of catalyzing oxidative phosphorylation or the P_i–ATP exchange unless the required coupling factor or factors were put back. Various apoparticles and their requirements are shown in Table 6-1. The coupling factors listed in parentheses indicate a partial and variable dependency.

Dependency of a particle preparation on a factor in a biological assay does not necessarily indicate a complete physical resolution. It is not surprising that this curious fact has led to confusion. For example, A-particles have only a minor depletion with respect to the mitochondrial ATPase yet they are virtually completely dependent on added F_1 in an assay of P_i–ATP exchange of or oxidative phosphorylation. Moreover, Lee and Ernster (1966) observed that in these particles low concentrations of oligomycin substituted for F_1, and they therefore suggested

TABLE 6-1 **Apoparticles and Coupling Factors from Mitochondria and Chloroplasts**

Apoparticles	Procedure of preparation	Reference	Factor dependency for P_i–ATP exchange
N-particles	Exposure of mitochondria to Nossal shaker	Penefsky *et al.* (1960)	F_1
SMP	Exposure of mitochondria to sonic oscillation at neutral pH	Racker (1962)	(F_1) (F_2)[a]
A-particles or EDTA-particles	Exposure of mitochondria to sonication at pH 9.2 in the presence of 0.5 m*M* EDTA	Fessenden and Racker (1966)	F_1 (F_2), OSCP
ASU-particles	Exposure of A-particles to urea after passage through Sephadex	Racker and Horstman (1967)	F_1 (F_2), OSCP (F_6)
STA-particles	Exposure of A-particles to silicotungstate	Racker *et al.* (1969)	F_1, F_2, OSCP, F_6
PC-particles (chloroplasts)	Exposure of Chloroplasts to sonication in the presence of P-lipids	Vambutas and Racker (1965)	CF_1
EDTA-particles (chloroplasts)	Exposure of chloroplasts to salt-free media containing 0.5 m*M* EDTA	Jagendorf and Smith (1962)	CF_1
STA-particles (chloroplasts)	Exposure of subchloroplast particles to silicotungstate	S. Lien and Racker (1971)	CF_1

[a]Factor in parenthesis indicates a small and variable stimulation.

that coupling factors do not have a catalytic function. On the other hand, when antibodies against F_1 inhibited oxidative phosphorylation (Fessenden and Racker, 1966), oligomycin had no stimulatory effect. Moreover, if particles were completely devoid of F_1 as in ASU-particles (Racker and Horstman, 1967), oligomycin did not stimulate unless some active F_1 was added as well. Similar observations were made with OSCP. Whereas A-particles require addition of OSCP for oxidative phosphorylation, the residual ATPase activity is very high and fully sensitive to oligomycin. The ability of low concentrations of oligomycin to substitute for OSCP as well as for F_1 suggests that both factors must have a structural role in the reconstitution of the membrane. F_2 (factor B) can be assayed in A-particles (Lam *et al.*, 1967), yet as pointed out earlier, either oligomycin or F_1 plus OSCP can substitute for factor B. Thus F_2 must also have a structural role in the proper assembly of the membrane. Only in STA-particles which are much more highly resolved particles, can we see a marked stimulation of phosphorylation by each of the individual coupling factors and much less effectiveness of oligomycin in stimulating phosphorylation (Racker *et al.*, 1969). It is therefore probable that like F_1 the other factors have also dual roles. I have pointed out in Lecture 4 that OSCP has a control function on ATPase activity, inhibiting ATP hydrolysis catalyzed by a phospholipid depleted complex.

ASU-Particles reconstituted with F_1, F_2, OSCP, and F_6 yield with DPNH as substrate P:O ratios between 2.5 to 3. With STA-particles which are stimulated two- to four-fold by F_2 and even more by F_6, the best P:O ratios we have obtained with DPNH are still below 2.

Similar to the observations made with apoenzymes, the stability of apoparticles decreases with the degree of resolution. The most resolved ASU- and STA-particles are stable only when kept in liquid nitrogen in the presence of 2 m*M* dithiothreitol, while SMP-particles are stable at $-70°$ without a thiol compound. Particles resolved with thiocyanate are even unstable at liquid nitrogen temperatures. Another variable feature is the time required for reconstitution. The more resolved the particles, the longer is the incubation time required for optimal reconstitution of the coupling factors.

Submitochondrial particles prepared by 2 minute sonication of mitochondria and resolved submitochondrial particles are loosely coupled, i.e., they respire without addition of P_i and ADP (Racker and Horstman, 1972). However, in contrast to uncoupled particles (e.g., in the presence of an uncoupler), they are capable of generating ATP provided an ATP trapping system such as glucose and hexokinase is added in sufficient amount to compete with the ATPase activity, and in the case of resolved particles sufficient coupling factors are added. Loosely coupled ASU-particles can be converted to tightly coupled particles by addition of an excess of coupling factors (Cockrell and Racker, 1969; Racker and Horstman, 1972).

Although resolved membrane preparations such as the ASU-particles have been primarily used as test particles for the assay of coupling factors, they are now being increasingly used for examination of the allotopic properties of membrane components. Thus, interesting studies have been performed on the properties of soluble ATPase and F_1 attached to ASU-particles. Significant differences in the kinetic properties and in adenine nucleotide binding have been observed (Hilborn and Hammes, 1973). ASU-Particles with and without F_1 have also served in comparative studies of the staining properties of F_1 in electron micrographs of thin sections of submitochondrial particles (Telford and Racker, 1973). STA-Particles prepared from chloroplasts have allowed the unambiguous identification of CF_1 in electron micrographs of freeze-etch preparations of reconstituted vesicles (Garber and Steponkus, 1974).

As was mentioned in Lecture 2, some of the procedures that were used for the preparation of resolved submitochondrial particles were also successfully used for the preparation of subchloroplast particles. However, urea treatment is not suitable for chloroplasts probably because the ATPase inhibitor is more firmly attached to CF_1 and protects the enzyme. The simplest procedure to prepare particles that are dependent for photophosphorylation on the addition of CF_1 is treatment of chloroplasts with EDTA as described in Lecture 2. However, similar to A-particles from mitochondria, EDTA chloroplasts are only partially depleted with respect to CF_1. The stimulation of photophosphorylation by inactivated CF_1 or even by DCCD points to a structural role of the chloroplast coupling factor. In order to obtain dependency on catalytically active CF_1, treatment with silicotungstate was required which unfortunately damaged the chloroplast membrane so that only relatively low rates of photophosphorylation were obtained after reconstitution (S. Lien and Racker, 1971).

Reconstitution of the Coupling Device

While an oligomycin-sensitive ATPase could be reconstituted by simply mixing hydrophobic components of mitochondria, phospholipids, and coupling factors, these preparations did not catalyze P_i–ATP exchange. After we learned that submitochondrial particles treated with cholate regained P_i–ATP exchange activity after dialysis (Arion and Racker, 1970), we exposed the oligomycin-sensitive ATPase to the same procedure and reconstituted the coupling device (Kagawa and Racker, 1971). As shown in Fig. 6-1 after 18 to 20 hours of dialysis, the preparation had maximal exchange activity. Before dialysis, CF_0 was amorphous; after reconstitution small vesicles were seen in electron micrographs. At this point it was important to establish that these vesicles were formed *de novo.* We found that radioactive inulin was trapped inside the vesicles only when added prior to reconstitution.

In earlier experiments, a crude mixture of soybean phospholipids was used in the reconstitution experiments. Later we found that pure phosphatidylethanolamine and phosphatidylcholine in 4:1 ratio was even better (Kagawa *et al.,* 1973a). When a 1:1 ratio (present in the natural membrane) was used, the rate of P_i–ATP exchange was lower but was markedly stimulated by addition of small amounts of cardiolipin. Large amounts of cardiolipin actually inhibited. Preparations of synthetic phospholipids were also active even when they contained unnatural side chains (Table

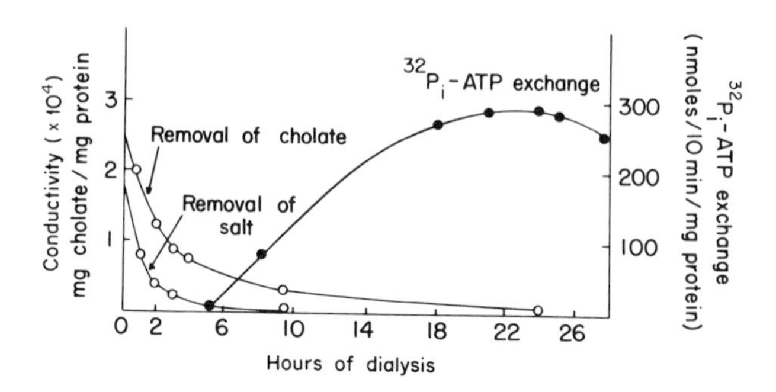

Fig. 6-1 Time course of reconstitution of P_i–ATP exchange. Experimental conditions were as described by Kagawa and Racker (1971).

6-2). With each new phospholipid introduced, particularly when it differed with respect to unsaturation, it was essential to establish the optimal ratio of the individual phospholipids. These experiments showed that some degree of unsaturation was required for the formation of active vesicles, but this conclusion was tempered by the observation that the saturated lipids were very difficult to disperse by sonication even in the presence of cholate. Thus, physical parameters governing vesicle formation enter into the considerations. The recent observations that an active proton pump with bacterial rhodopsin can be reconstituted with fully saturated phospholipids, provided sonication is performed at temperatures above the transition temperature (Racker and Hinkle, 1974), lends support to this view.

The vesicles reconstituted by the cholate dialysis procedure catalyzed a rapid P_i–ATP exchange and resembled submitochondrial particles in their response to 1-anilino-8-naphthalene sulfonate. On addition of ATP a marked enhancement of fluorescence was noted. This method of evaluating the "energization" of the membrane proved most useful in the reconstitution of oxidative phosphorylation. The most important property of the reconstituted vesicles for our understanding of oxidative phosphorylation is that they catalyze an ATP-dependent translocation of protons. These vesicles represent the first reconstitution of a

TABLE 6-2 **Reconstitution of Active Vesicles with Pure Phospholipids**[a]

Phospholipids	$^{32}P_i$–ATP exchange (nmoles $AT^{32}P$/ min/mg protein)
Exp. 1	
soy phosphatidylcholine (PC)	0.3
soy phosphatidylethanolamine (PE)	8
PE + PC (4:1)	28
PE + PC (2.5:2.5)	22
PE + PC (2.5:2.5) + cardiolipin (0.15)	50
Exp. 2	
diphytanoyl PC	5
1-stearoyl-2-undecenoyl PE	0
diphytanoyl PC + 1-stearoyl-2-undecenoyl PE (2:3)	24

[a]The experimental conditions were as described by Kagawa *et al.* (1973a). The values in parentheses give the ratios of the phospholipids (5 μmoles total).

proton pump (Kagawa *et al.,* 1973a). As shown in Fig. 6-2 at pH 6.25 in the presence of valinomycin and K^+, there was an uptake of protons on addition of ATP, which was abolished by uncouplers. The procedure of measuring ATP-driven proton translocation at pH 6.25 was developed by Thayer and Hinkle (1973a) to avoid changes in pH caused by the hydrolysis of ATP.

In the case of the reconstitution of the chloroplast coupling device a procedure similar to that applied to the mitochondrial system has been used (Carmeli and Racker, 1973). The cholate dialysis procedure yielded vesicles that catalyze an uncoupler and DCCD-sensitive P_i–ATP exchange in the presence of dithiothreitol. The specific antibody against CF_1 inhibited the exchange.

The cholate dialysis procedure of reconstitution described above yields very active vesicles, but has the inconveniences associated with the prolonged dialysis. A more rapid reconstitution procedure was developed (Racker, 1973) which consists of

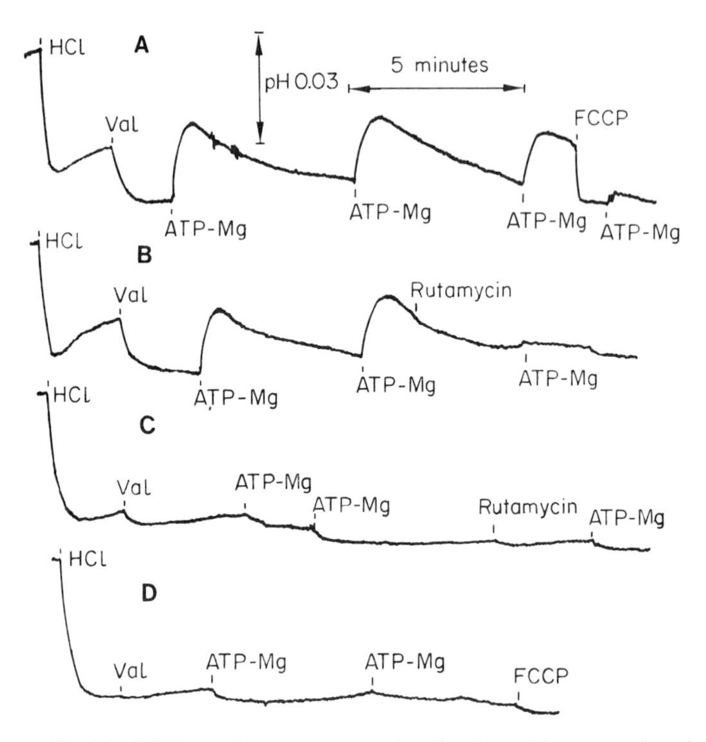

Fig. 6-2 ATP-Dependent proton translocation in vesicles reconstituted with oligomycin-sensitive ATPase. Experimental conditions were as described by Kagawa *et al.* (1973a). A, Vesicles reconstituted with a crude mixture of soybean phospholipids; B, vesicles reconstituted with purified phosphatidylcholine and phosphatidylethanolamine; C, vesicles reconstituted with phosphatidylcholine; D, vesicles reconstituted with phosphatidylethanolamine.

exposing phospholipids and membranous proteins suspended in buffer to sonic oscillation. In the case of the P_i–ATP exchange the rates obtained with vesicles prepared by the sonication procedure are usually about half of those obtained by the cholate dialysis procedure (Racker, 1975a). On the other hand, in some other reconstitutions such as the rhodopsin proton pump and the Na^+-K^+ pump, sonication is far superior to cholate dialysis. The most rapid and convenient method for the reconstitution of the

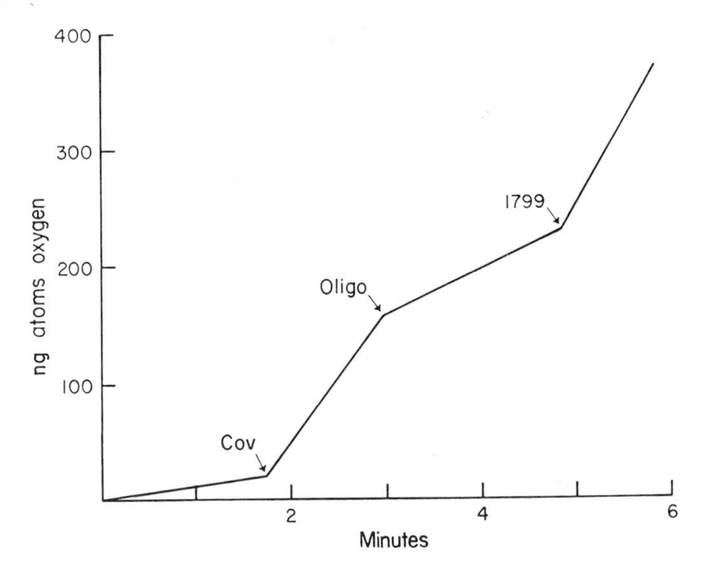

Fig. 6-3 Loss of respiratory control by CF_0. Experimental conditions were as described by Racker (1972b). Cytochrome oxidase vesicles (COV) were reconstituted in the presence of the mitochondrial hydrophobic protein (without added coupling factors). Respiration was tested with cytochrome *c* and ascorbate, then oligomycin was added, then an uncoupler (1799).

P_i–ATP exchange is the cholate dilution procedure (Racker *et al.*, 1975b). It consists of exposing the phospholipids and proteins to about 0.7% cholate and diluting at least twentyfold for assay. Another procedure that is also effective in the reconstitution of the coupling device is the incorporation procedure (Eytan *et al.*, 1975b), in which the membrane protein is incorporated by simply mixing with liposomes containing lysolecithin (10% of the total phospholipid) or acidic phospholipids (Eytan and Racker, 1976).

Addition of coupling factors to CF_0 preparations or to apoparticles serves as suitable assay for the components that can be isolated as water-soluble proteins. In order to test the hydrophobic CF_0 independently of coupling factors, we have taken advantage of the capacity of CF_0 to serve as an oligomycin- or

DCCD-sensitive hydrogen channel (Racker, 1972b). In this system which has been used as an assay for CF_0 preparation from either mitochondria or chloroplasts, the respiratory control of cytochrome oxidase vesicles (see below) is abolished by the presence of the hydrogen ion channel. As shown in Fig. 6-3 sensitivity to an energy transfer inhibitor is monitored to ensure that we are measuring specific hydrogen ion movements and not those caused by leakage of the membrane due to insertion of denatured hydrophobic proteins.

Reconstitution of the Mitochondrial Oxidation Chain with Respiratory Control

Segments of the oxidation chain have been incorporated into phospholipid vesicles by the same cholate dialysis procedure which we have used to reconstitute the mitochondrial proton pump. The first system with respiratory control was cytochrome oxidase vesicles (Hinkle *et al.*, 1972). These vesicles catalyzed the oxidation of reduced cytochrome *c* at a very low rate which was increased five- to tenfold when K^+, valinomycin, and nigericin were added. Valinomycin in the presence of K^+ collapsed the membrane potential; nigericin, which catalyzes a K^+/H^+ exchange, collapsed the proton gradient. As shown in Table 6-3 neither of the two ionophores alone was effective, as predicted by the

TABLE 6-3 **Respiratory Control in Cytochrome Oxidase Vesicles**[a]

	ng atoms oxygen/min	Respiratory control ratio
Cytochrome oxidase vesicles	28	
+ 0.2 μg valinomycin	30	1.07
+ 0.2 μg nigericin	54	2.0
+ 40 μg oleic acid	61	2.1
+ valinomycin + nigericin	205	7.3
+ valinomycin + oleic acid	188	6.7

[a]Experimental procedure was as described by Racker (1972b).

chemiosmotic hypothesis. It can also be seen from this table that fatty acids act as proton ionophores similar to nigericin. There was only little effect on respiration in the absence of valinomycin. More recently, the first and second segment of oxidation was incorporated into liposomes by the same procedure and both were shown to exhibit respiratory control (Hinkle and Leung, 1974; Ragan and Hinkle, 1975).

Reconstitution of Oxidative Phosphorylation

The oxidation chain which we discussed in the last lecture can be reconstituted either from the complexes (Hatefi *et al.*, 1962) or from further resolved components (Yamashita and Racker, 1968). Numerous attempts in various laboratories to restore phosphorylating activity to such systems have until recently been unsuccessful. After we succeeded in the reconstitution of the uncoupler-sensitive $^{32}P_i$–ATP exchange (Kagawa and Racker, 1971), we thought that reconstitution of oxidative phosphorylation would be a simple matter. Yet, in the beginning all experiments failed until we realized the importance of the asymmetric reconstitution.

I shall now describe the successful reconstitution of the third site of oxidative phosphorylation and then come back to the reason for our earlier failures. The reconstitution of the third site of oxidative phosphorylation required cytochrome oxidase, cytochrome *c*, the mitochondrial proton pump, and phospholipids. We first used a rather crude mixture of soybean phospholipids that were suspended in 2% cholate and clarified by sonication. The lipids and proteins were mixed and dialyzed overnight against a suitable buffer. After addition of bovine serum albumin and $MgCl_2$ the vesicles were sedimented through a layer of sucrose to remove, as effectively as possible, external cytochrome *c*. After incubation with coupling factors the vesicles catalyzed oxidative phosphorylation with ascorbate-phenazine methosulfate as substrate (Racker and Kandrach, 1973).

It may be instructive to relate the events which led to this successful reconstitution. First, we decided to use two assays to evaluate the success of reconstitution. The first was oxidative phosphorylation. The second was a rapid fluorescence assay for site 3 based on the observations of Azzi *et al.* (1969) that the fluorescence of ANS is greatly enhanced by an energized membrane. These authors showed that energization can be achieved by substrate oxidation or by ATP. This allowed us to evaluate separately the reconstitution of the coupling device (membrane energization by ATP) and of the oxidation chain (membrane energization by PMS-ascorbate). When the dialyzed vesicles were first tested there was no indication that they catalyzed oxidative phosphorylation. The ANS assay showed that ATP energized the membrane but not PMS-ascorbate. In fact the results were more than negative, since a slight response was noted in the opposite direction, namely, a quenching of fluorescence. Since a decrease of fluorescence is characteristic of the response of mitochondria, I asked Dr. Peter Hinkle to check on the polarity of our vesicles by measuring proton movements. He observed movements characteristic for mitochondrial rather than submitochondrial particles. It thus became clear to us that we had preferentially reconstituted the mitochondrial rather than the submitochondrial conformation and that it was likely that the lack of asymmetry was responsible for our failure to observe oxidative phosphorylation. We then used a procedure developed for the removal of external cytochrome *c* from submitochondrial particles (Arion and Wright, 1970). Furthermore, we added polylysine to our assay to displace any residual external cytochrome *c*. Under these conditions, we finally observed respiration-dependent enhancement of ANS fluorescence (Fig. 6-4) and oxidative phosphorylation in the reconstituted system with P:O ratios of 0.5 or higher (Racker and Kandrach, 1971, 1973).

One conclusion that can be drawn from these experiments which is relevant for future studies is that reconstitution by the cholate dialysis procedure gives rise to the assembly of cytochrome oxidase in both directions with the mitochondrial conformation (cytochrome *a* on the outside) predominating. Thus, in contrast to native submitochondrial particles, the reconstituted vesicles are readily uncoupled by externally added cytochrome *c*

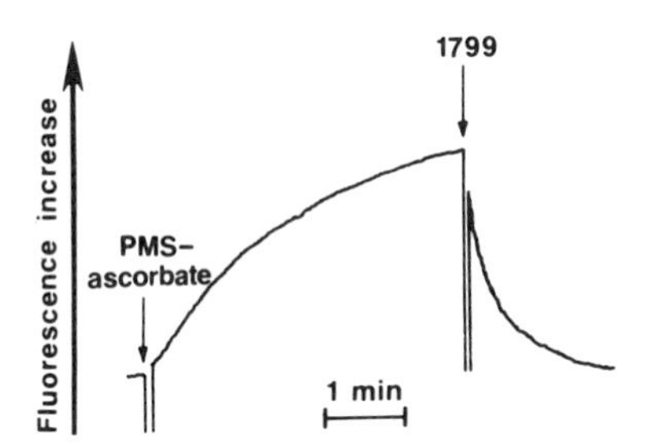

Fig. 6-4 Respiration-dependent fluorescence enhancement in reconstituted vesicles. Experimental conditions were as described by Racker and Kandrach (1971). Vesicles were reconstituted with cytochrome oxidase, cytochrome *c*, and the oligomycin-sensitive ATPase of mitochondria. Fluorescence was measured in an Eppendorf fluorimeter.

as shown in Table 6-4. This uncoupling by cytochrome *c* can be readily understood in terms of the chemiosmotic hypothesis, since cytochrome oxidase operative in both directions would eliminate the proton motive force. The uncoupling by cytochrome *c* is much more difficult to explain within the framework of the chemical hypothesis. We have recently developed reconstitution procedures that are more selective than the cholate dialysis procedure and give rise to unidirectional assembly of protein complexes (Eytan *et al.*, 1976).

Later, we substituted for the crude phospholipid mixture, purified phosphatidylethanolamine and phosphatidylcholine. The optimal ratio varied with the source of the phospholipids. With soybean phospholipids it was 4:1 as in the case of the coupling device. With a 1:1 ratio phosphorylation was quite low, but was again markedly increased by small amounts of cardiolipin (Table 6-5).

By an identical procedure site 1 was reconstituted. With complex I of Hatefi *et al.* (1962) and a 4:1 mixture of phosphatidylethanolamine and phosphatidylcholine, we reconstituted vesicles that catalyzed oxidative phosphorylation after addition of coupling factors (Ragan and Racker, 1973a). To remove excess of complex I which was not incorporated, the vesicles were separated by centrifugation in a sucrose gradient; P:O ratios of about

TABLE 6-4 **Uncoupling of Oxidative Phosphorylation by Cytochrome *c*[a]**

Additions	O_2 uptake (atoms/min)	ATP formation (nmoles/min)	P:O ratio
Complete reconstituted system	25.0	9.3	0.37
– polylysine	38.3	5.3	0.17
– polylysine + cytochrome *c* (10 μg)	55.0	2.7	0.07
– polylysine + cytochrome *c* (25 μg)	60.0	1.8	0.04
+ ETP_H particles	25.0	18.8	0.76
+ ETP_H particles + cytochrome *c* (25 μg)	49.0	19.4	0.54

[a]The reconstituted vesicles (350 μg) were assayed as described by Racker and Kandrach (1973) with and without 50 μg of polylysine. The ETP_H particles were assayed only without polylysine.

TABLE 6-5 **Reconstitution of Site 3 with Purified Phospholipids**[a]

Phosphatidylethanolamine	:	phosphatidylcholine	:	cardiolipin	P:O
4	:	1	:	0	0.38
2.5	:	2.5	:	0	0.11
2.5	:	2.5	:	0.15	0.33

[a]Experimental procedure was as described by Racker and Kandrach (1973).

0.5 were obtained. In this case we did not have to be concerned with the asymmetric organization of the membrane since DPNH did not penetrate into the inside of the vesicles. Actually, estimates from the bleaching of the flavin by externally added DPNH in the presence and absence of detergents suggested that the reconstitution was random, about 50% in the mitochondrial and 50% in the submitochondrial orientation. As mentioned in the last lecture, we have resolved from complex I a rotenone-sensitive DPNH dehydrogenase. As shown in Table 6-6, a high P:O ratio was observed when this preparation was used in reconstitution (Ragan and Racker, 1973b).

TABLE 6-6 **Reconstitution of Site 1 Phosphorylation with DPNH Dehydrogenase**[a]

Additions	nmoles DPNH	nmoles $AT^{32}P$	P:2*e*
Complex I (reconstituted)	79.4	40.4	0.66
+ rotenone	18.4	0.66	
DPNH dehydrogenase (reconstituted)	61.8	14.5	0.54
+ rotenone	35.1	0	
+ 1799	62.5	0	
+ rutamycin	61.8	0	

[a]Experimental procedure was as described by Ragan and Racker (1973b). P:2*e* ratios were determined after correction for rotenone-insensitive oxidation.

TABLE 6-7 **Reconstitution of Vesicles Catalyzing Phosphorylation Coupled to the Oxidation of Succinate**[a]

Additions	Oxidative phosphorylation		
	ng atoms O	nmoles glucose-6-P	P:O
Succinate	64	40	0.63
PMS-Ascorbate	184	64	0.35

[a]Experimental conditions were as described by Racker *et al.* (1975b).

We have recently reconstituted site 2 plus site 3 of oxidative phosphorylation (Racker *et al.,* 1975b). Reconstitutions with complex III (Hatefi *et al.,* 1962), cytochrome oxidase, cytochrome *c*, and phosphatidylethanolamine and phosphatidylcholine (4:1) from mitochondria either by the cholate dialysis or by the cholate dilution procedure yielded vesicles which, after further addition of succinate dehydrogenase and coupling factors, catalyzed oxidative phosphorylation. With succinate the P:O ratios were about twice as high as the P:O ratio obtained in the same particles with ascorbate-phenazine methosulfate as substrate (Table 6-7). Thus all three sites of oxidative phosphorylation can be reconstituted into artificial liposomes.

Reconstitution of the Proton Pump of *Halobacterium halobium* and of Rhodopsin-Catalyzed Photophosphorylation

It was shown by Oesterhelt and Stoeckenius (1973) that *Halobacterium halobium* bacteria contain rhodopsin in purple patches that respond to light activation by translocation of protons from the inside to the outside of intact bacteria. Danon and Stoeckenius (1974) observed increases in the intracellular ATP levels associated with the light-driven proton translocation. Vesicles were reconstituted with bacteriorhodopsin and phospholipids

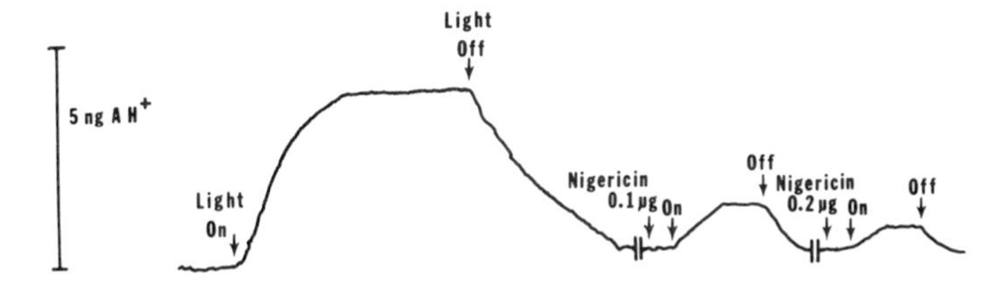

Fig. 6-5 Proton uptake in vesicles reconstituted with bacteriorhodopsin. Experimental conditions were as described by Racker and Hinkle (1974). Vesicles were reconstituted with soybean phospholipids and bacteriorhodopsin by the sonication procedure (Racker, 1973).

either by the cholate dialysis procedure (Racker and Stoeckenius, 1974) or by the sonication procedure (Racker, 1973). They catalyzed the translocation of protons from the outside to the inside, in the opposite direction from that of the intact bacteria as shown in Fig. 6-5. The purple membrane preparation used in these experiments contained rhodopsin as the only polypeptide chain as well as about 25% of phospholipids. In this experiment a single synthetic phospholipid (dimyristoyl phosphatidylcholine) was used to prepare the vesicles. As can be seen from Fig. 6-5 nigericin, an ionophore which catalyzes an H^+/K^+ exchange, collapsed the light-dependent pH gradient at low concentrations.

By sonic oscillation of *H. halobium* we have obtained two types of subbacterial particles (Kanner and Racker, 1975). The larger particles catalyzed light-dependent proton translocation from the inside to the outside like the intact bacteria. They also catalyzed a valinomycin-stimulated and light-dependent uptake of $^{86}Rb^+$. The small particles were inside-out and catalyzed proton translocation from the outside to the inside like the reconstituted vesicles. It seems likely that in this case the size of the vesicles is a determinant in the orientation of the rhodopsin molecules.

When bacteriorhodopsin was reconstituted together with the mitochondrial oligomycin-sensitive ATPase, light-dependent formation of ATP was observed as shown in Table 6-8. In the absence of light there was no formation of radioactive ATP. This is an important control which shows that the P_i–ATP exchange

TABLE 6-8 **ATP Formation by Mitochondrial Proton Pump Reconstituted with Bacteriorhodopsin**

Additions	nmoles ATP	nmoles ATP/mg bacteriorhodopsin
Complete system (illuminated)	15.5	594
minus coupling factors	2.9	110
plus 1799 (10^{-4} *M*)	0	0
plus rutamycin (4 μg)	4.6	170
(dark)	0.6	23

[a]Experimental procedure was as described by Racker and Stoeckenius (1974). The complete system contained the liposomes reconstituted with bacteriorhodopsin and the oligomycin-sensitive ATPase from mitochondria. The reconstituted liposomes were incubated with F_1 and OSCP prior to assay.

can be effectively surpressed in the presence of excess hexokinase and does not interfere with the analysis of net ATP formation.

The reconstitution of oxidative phosphorylation by the combination of the respiratory chain with the mitochondrial proton pump and particularly the demonstration that the bacterial proton pump can substitute for the respiratory chain strongly suggest that the major function of the respiratory chain is indeed the translocation of protons as formulated by the chemiosmotic hypothesis.

I therefore believe that one of the major questions of oxidative phosphorylation, namely, the coupling of oxidation to phosphorylation, has been solved by the formulations of the chemiosmotic hypothesis. But a second and equally challenging problem is still ahead of us. How does a pump utilize a proton flux to generate ATP? This will be the subject of the next lecture.

Lecture 7

Reconstitution and Mechanism of Action of Ion Pumps

Prejudice is a disease characterized by hardening of the categories.

W. A. Ward

Criticism may not be agreeable, but it is necessary. It fulfills the same function as pain in the human body: it calls attention to an unhealthy state of things.

Winston Churchill

General Comments

When I became convinced that the basic principle of the coupling mechanism between oxidation and phosphorylation had been solved by the formulations of the chemiosmotic hypothesis, I felt that I wanted to turn my major attention to the next problem. How can an ion gradient be utilized to generate ATP from ADP and P_i? How is ATP utilized to generate an ion gradient?

Faced with this problem, I decided to expand our research to ion pumps other than the proton pump of mitochondria. The major reason for this decision was the complexity of the mitochondrial pump with its hydrophobic components and multiple coupling factors (see Lecture 4). In contrast, a Ca^{2+}-ATPase complex has been prepared from sarcoplasmic reticulum (MacLennan, 1970) that contained only one major protein component and a Na^+-K^+-ATPase complex has been isolated from the plasma membrane which contained only two components (Kyte, 1971). Both the Ca^{2+} pump of sarcoplasmic reticulum and the Na^+-K^+ pump of the plasma membrane have been shown to generate ATP when an ion gradient was dissipated on reversal of the pump action (Glynn and Lew, 1969; Makinose and Hasselbach, 1971; Panet and Selinger, 1972). Moreover, with both ATPase preparations a phosphorylated intermediate was demonstrated (Post *et al.,* 1965; Martonosi, 1967; Fahn *et al.,* 1968; Yamamoto and Tonomura, 1968). Although there was ample indirect evidence that these ATPases are functional parts of the pumps, we did not know whether the purified ATPase is all that is needed for ion translocation, or whether, as in the case of mitochondria, additional coupling factors are required. We found through reconstitution experiments that highly purified preparations of the Ca^{2+}-ATPase can be incorporated into liposomes which catalyzed an ATP-dependent translocation of Ca^{2+} (Racker, 1972b). I will show later that in fact these highly purified preparations contain an additional component that is required for Ca^{2+} translocation. Nevertheless, it is quite apparent that this pump is much simpler than the mitochondrial proton pump. Moreover, the sarcoplasmic reticulum is a much simpler organelle so that complications caused by side reactions which have plagued mitochondrial research are not as likely to occur.

Reconstitution of Ion Pumps

The Ca^{2+} Pump of Sarcoplasmic Reticulum

The reconstitution of this ion pump has been particularly easy. It can be achieved by the cholate dialysis procedure (Racker, 1972a), by the sonication procedure (Racker and Eytan, 1973), or by the cholate dilution procedure (Racker *et al.,* 1975b).

As shown in Table 7-1 the ATP-driven Ca^{2+} translocation in vesicles reconstituted by the sonication procedure was sensitive to ionophores for divalent cations such as A-23187. Valinomycin in

TABLE 7-1 **Reconstitution of Calcium Pump by the Sonication Procedure without Detergent**[a]

Assay conditions	ng ions Ca^{2+}/min/mg protein
Complete	181
without ATP-Mg	15
without ATP	8
plus A-23187	4
plus valinomycin (2 μg)	213
plus valinomycin + nigericin (2 μg)	218

[a]Experimental conditions were as described by Racker and Eytan (1973).

the presence of K^+ slightly but consistently increased the rate of Ca^{2+} uptake, suggesting that a membrane potential may be formed during translocation.

In the course of these studies we observed considerable variations in the reconstitutive capacity of different ATPase preparations. Some fractions with a low specific activity of ATPase were more active in Ca^{2+} transport than some other fractions with much higher ATPase activity (Table 7-2). Over tenfold differences were observed in Ca^{2+} transport in preparations with the same specific activity for ATPase. It should be noted that the ATP hydrolysis during Ca^{2+} translocation in the reconstituted system varied greatly dependent on the fraction used. As shown in Table 7-3, a heated extract of an active preparation with low ATPase activity markedly stimulated transport activity of a preparation with low transport activity (Racker and Eytan, 1975). An analysis of the various fractions by acrylamide gel electrophoresis revealed that reconstitutively active preparations contained a much greater amount of a polypeptide band with an R_f corresponding to that of the proteolipid described by MacLennan *et al.* (1972). This is shown in Fig. 7-1. The heated "coupling factor" prepared from the active fraction contained this proteolipid as the major component. We then prepared the proteolipid by chloroform–methanol extraction and precipitation by ether as

TABLE 7-2 **Comparison of ATPase Activity and Ca^{2+} Translocation**

		Reconstituted vesicles[b]		
Isolated fraction	ATPase[a] (μmoles/min/mg protein)	Ca^{2+} translocation (ng ions/min/mg protein)	ATP hydrolysis (nmoles/min/mg protein)	Ca^{2+}/ATP ratio
R_{3a}	5.6	50	160	0.31
R_{3b}	8.0	242	630	0.38
R_{3c}	11.0	365	930	0.39
R_{3d}	8.6	646	708	0.91
R_{3e}	5.0	826	490	1.7

[a]The ATPase activity of the isolated fractions was assayed at 37° (MacLennan, 1970).

[b]The Ca^{2+} transport activity and the ATP hydrolysis during Ca^{2+} translocation was measured after 5 min incubation at room temperature as described by Racker and Eytan (1975).

TABLE 7-3 **Stimulation of Ca^{2+} Transport by Heat-Stable Coupling Factor**[a]

	Ca^{2+} translocation (ng ions/min/mg protein)	ATP hydrolysis (nmoles/min/mg protein)	Ca^{2+}/ATP ratio
Exp. 1			
R_{3a}	50	160	0.31
R_{3a} + 3 μg coupling factor	144	180	0.80
R_{3a} + 6 μg coupling factor	180	130	1.40
Exp. 2			
R_{3a}	190	730	0.26
R_{3a} + 4.5 μg coupling factor	370	510	0.72
R_{3a} + 15 μg coupling factor	120	1010	0.12

[a] Experimental conditions were as described by Racker and Eytan (1975).

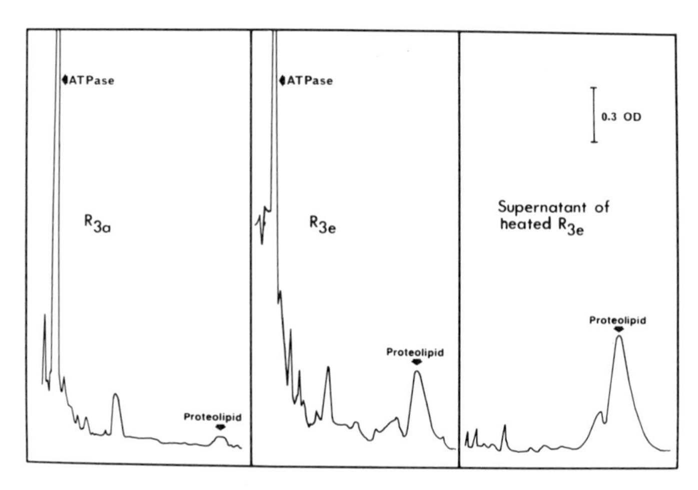

Fig. 7-1 Acrylamide gel scans of different preparations of Ca^{2+}-ATPase and of the heat-stable coupling factor. Experimental conditions were as described by Racker and Eytan (1975). The Ca^{2+}-ATPase was purified according to MacLennan (1970) by precipitation with ammonium acetate. R_{3a} was the first precipitate, R_{3e} the last precipitate.

described by MacLennan *et al.* (1972), and found that it was also active as a coupling factor when tested immediately after isolation. It was, however, less active than the heated water-soluble coupling factor. In contrast to the latter, which was stable for months in the refrigerator, the solvent-extracted proteolipid became insoluble on storage and one day following its isolation was inactive as a coupling factor. We were therefore not surprised to find that the preparation of proteolipid, kindly sent to us by Dr. MacLennan, was inactive. I would like to draw your attention to the marked increase in the Ca^{2+}/ATP ratio indicating an improvement of the efficiency of pump operation by the proteolipid (Table 7-2). We shall return to this observation later.

Since this was the first indication for a function of a proteolipid in a biological pump, it seems appropriate to review briefly the properties of these curious biopolymers, particularly since none of the biochemistry textbooks mentions them. The first isolation of proteolipids was from brain (Folch and Lees, 1951).

Later, proteolipids were isolated from other animal tissues, plants, and microorganisms (cf. Folch-Pi and Stoffyn, 1972). They are proteins that are soluble in chloroform–methanol (2:1) and contain covalently bonded fatty acids but no phospholipids. They are remarkably resistant to the action of proteases. They are very hydrophobic when prepared by solvent extraction, but can be converted into hydrophilic proteins by a process of slow substitution of water for the organic solvent. Aqueous solutions containing 2% protein or more can thus be prepared.

We observed that reconstitution of the Ca^{2+} pump with the proteolipid required careful titration. When too much of the latter was added the transport activity was completely abolished. This led us to the notion that an excess of the proteolipid may act as an ionophore similar to the action of A-23187. This was shown to be the case (Table 7-4). This was reminiscent of the

TABLE 7-4 **Effect of Factor Added to Reconstituted Systems Catalyzing $^{45}Ca^{2+}$ Transport and to $^{45}Ca^{2+}$-Loaded Liposomes[a]**

Liposomes	Additions during assay	$^{45}Ca^{2+}$ translocation (ng ions/min/mg protein)	cpm
Exp. 1			
Liposomes reconstituted with ATPase	None	462	
	10 μl (3 μg) of factor	392	
	25 μl (7.5 μg) of factor	219	
	50 μl (15 μg) of factor	132	
	50 μl of elution buffer	437	
Exp. 2			
Liposomes loaded with $^{45}CaCl_2$	None		22,000
	25 μl (7.5 μg) of factor		17,000
	50 μl (15 μg) of factor		11,000
	50 μl of elution buffer		19,800

[a]Experimental conditions were as described by Racker and Eytan (1975). The factor (proteolipid) was added shortly prior to assay rather than during reconstitution.

TABLE 7-5 **Effect of Mitochondrial and Chloroplast Proteolipids on Bacteriorhodopsin Proton Pump**[a]

Additions	Proton uptake (ng ions/mg protein)	
	+ valinomycin	– valinomycin
Exp. 1		
Bacteriorhodopsin vesicles	408	300
+ 10 μg mitochondrial proteolipid	360	–
+ 20 μg mitochondrial proteolipid	275	–
+ 30 μg mitochondrial proteolipid	192	162
+ 40 μg mitochondrial proteolipid	132	–
+ 50 μg mitochondrial proteolipid	67	115
+ FCCP (10^{-6} *M*)	32	97
Exp. 2		
Bacteriorhodopsin vesicles	411	
+ 10 μg chloroplast proteolipid	356	
+ 25 μg chloroplast proteolipid	213	
+ 35 μg chloroplast proteolipid	116	

[a]Experimental conditions were as described by Racker (1975b).

TABLE 7-6 **Effect of Mitochondrial and Chloroplast Proteolipids on Cytochrome Oxidase Vesicles**[a]

Additions	ng atoms oxygen/ min/μg
Cytochrome oxidase vesicles	5.5
+ 25 μg mitochondrial proteolipid	9.9
+ 50 μg mitochondrial proteolipid	13.6
+ 75 μg mitochondrial proteolipid	18.3
+ 25 μg chloroplast proteolipid	11.2
+ 50 μg chloroplast proteolipid	16.9
+ 75 μg chloroplast proteolipid	19.5
+ 0.3 μg valinomycin + 10^{-6} *M* FCCP	25.0

[a]Experimental conditions were as described by Racker (1975b).

TABLE 7-7 **Effect of Excess Phospholipid on Uncoupling by Proteolipid and FCCP**[a]

Additions	Protein uptake (ng ions/mg bacteriorhodopsin)
Exp. 1	
Bacteriorhodopsin vesicles	540
+ 25 μg proteolipid	374
+ 50 μg proteolipid	114
+ 50 μg proteolipid + 1 μmole phosphatidylethanolamine	104
+ 50 μg proteolipid + 1.4 μmoles phosphatidylethanolamine	102
Exp. 2	
Bacteriorhodopsin vesicles	485
+ FCCP (2.5 × 10^7 *M*)	135
+ FCCP (2.5 × 10^7 *M*) + 1 μmole phosphatidylethanolamine	375
+ FCCP (2.5 × 10^7 *M*) + 1.4 μmoles phosphatidylethanolamine	445

[a]Experimental conditions were as described by Racker (1975b).

observation that the hydrophobic protein fraction of mitochondria (CF_0), required for oxidative phosphorylation, inhibited when added in excess (Racker and Kandrach, 1973). Since CF_0 contains also the DCCD-binding proteolipid (Cattell *et al.*, 1970), I have tested proteolipid preparations from mitochondria and chloroplasts for proton ionophore activity. As shown in Table 7-5 both preparations collapsed the proton gradient generated by the reconstituted bacteriorhodopsin pump. They also stimulated respiration of tightly coupled cytochrome oxidase vesicles (Table 7-6). However, in contrast to the effect of highly mobile uncouplers such as FCCP, the effect of the proteolipids was not readily reversed by addition of excess phospholipids to the assay system (Table 7-7). I shall return to the function of the proteolipid in the discussion of the mechanism of action of the Ca^{2+} pump.

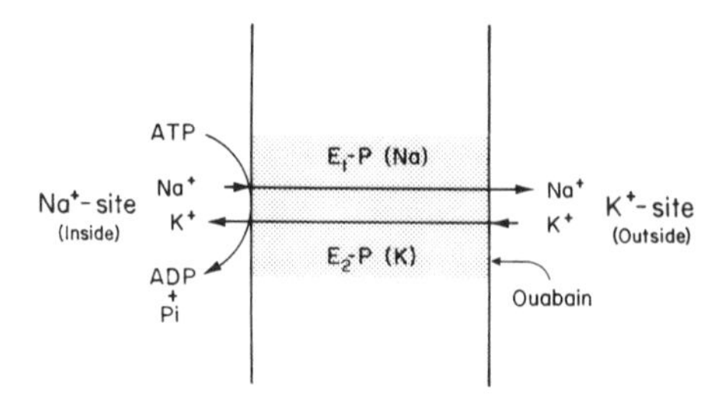

Fig. 7-2 Topography of the Na^+-K^+ pump of the plasma membrane.

The Na^+ Pump of the Plasma Membrane

Whereas the Ca^{2+} pump is normally activated by ATP from the outside of sarcoplasmic reticulum vesicles, the Na^+ pump of the plasma membrane operates with ATP on the inside of the cellular plasma membrane. To reconstitute a Na^+ pump there are two alternatives. As illustrated in Fig. 7-2 one can either prepare vesicles that are inside out and could therefore catalyze the uptake of Na^+ by external ATP and release of K^+, or one can prepare right side out vesicles that require internal ATP and which take up K^+ and release Na^+. As shown in Fig. 7-3 such a system could be reconstituted by including a transporter for

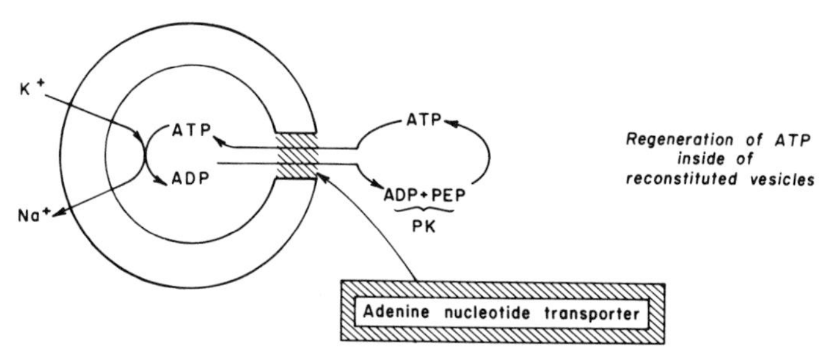

Fig. 7-3 A model for intravesicular regeneration of ATP. An ATP-regenerating system consisting of pyruvate kinase and phosphoenolpyruvate is used outside the vesicles. The adenine nucleotide transporter, by exchanging ADP for ATP, serves as an ATP regenerating system inside the vesicles.

adenine nucleotide, which is required also for the reconstitution of oxidative phosphorylation with a mitochondrial conformation of the membrane. To perform such experiments we have been engaged in the isolation of the mitochondrial adenine nucleotide transport carrier (Shertzer and Racker, 1974) and shall attempt reconstitutions of the type described in Fig. 7-3. Meanwhile we have reconstituted the Na^+ pump in inside out vesicles. Goldin and Tong (1974) and Hilden *et al.* (1974) have reconstituted such vesicles with purified Na^+-K^+-ATPase from dog kidney or from rectal gland of the dog fish using the cholate dialysis procedure. We have used a purified preparation of ATPase from electric eel for the same purpose (Albers *et al.,* 1963) and reconstituted active vesicles by the sonication procedure (Racker and Fisher, 1975). As shown in Table 7-8 the ATP-dependent Na^+ uptake was

TABLE 7-8 **Effect of Ionophores and Inhibitors on the Reconstituted Na^+ Pump[a]**

	$^{22}Na^+$ translocation (ng ions/min/mg protein)		
Additions	– ATP	+ ATP	Δ
Exp. 1			
None	177	439	262
+ 0.2 μg valinomycin	220	586	366
+ 10 μg rutamycin	233	415	182
+ 2 m*M* ouabain (external)	248	443	195
Exp. 2			
None	180	436	256
+ 0.2 μg valinomycin	245	553	308
+ 5 μg gramicidin (internal)	360	360	0
+ 2 m*M* ouabain (internal)	178	159	0

[a]Reconstitution with purified Na^+-K^+-ATPase from electric eel and assays were performed as described by Racker and Fisher (1975) and Gasko *et al.* (1976). In experiment 1, valinomycin and the inhibitors were added to the assay mixture. In experiment 2, gramicidin and ouabain were added prior to sonication.

TABLE 7-9 Effect of Phosphulipid Composition on Na^+ Translocation in Reconstituted Liposomes[a]

Phospholipid composition	$^{22}Na^+$ translocation (ng ions/min/mg protein) – ATP	+ ATP	Δ
Phosphatidylcholine (PC)	55	98	43
PC:PE (1:1)	56	130	74
PC:PE (1:2)	69	141	72
PC:PE (1:3)	76	187	111
PC:PE (1:4)	72	226	154
Phosphatidylethanolamine (PE)	95	295	200

[a]Experimental conditions were as described by Racker and Fisher (1975).

stimulated by valinomycin under conditions of equimolar concentrations of K^+ inside and outside the vesicles. This suggests that the translocation of Na^+ may be electrogenic in the reconstituted vesicles. Gramicidin or nigericin stimulated the energy independent ion movement but inhibited the ATP-dependent Na^+ translocation. Ouabain added during reconstitution inhibited. With purified phospholipids the highest Na^+ transport activity was observed with phosphatidylethanolamine (Table 7-9). In some experiments the presence of small amounts of phosphatidylcholine (25% of total phospholipids) slightly increased pump activity.

What Can We Learn from Resolution and Reconstitution Experiments?

Asymmetry and Orientation

We have learned that asymmetry of assembly is essential for the function of oxidative phosphorylation as illustrated by the pronounced uncoupling by cytochrome *c* in reconstituted vesi-

TABLE 7-10 **Orientation of Membrane Proteins in Reconstituted Vesicles**

	Oxidative phosphorylation site 1		Cytochrome oxidase				ADP–ATP transporter	
	+ HP	– HP	+ cytochrome *c*	– cytochrome *c*	Ca^{2+} pump	Bacteriorhodopsin proton pump	ETPH	Reconstituted
Right side in	50	80	60	100	100	0	0	100
Inside out	50	20	40	0	0	100	100	0

cles. The need for asymmetry seems obvious in systems that require ion translocation. But how does nature achieve proper orientation of the catalysts? If, as I have suggested earlier, a natural detergent such as lysolecithin is responsible for the incorporation of membrane components located on the matrix side but manufactured outside of mitochondria (cf. Schatz and Mason, 1974), some guidance for proper assembly seems required. This may involve an actual template, a "structural protein," or it may be a simple consequence of the orientation of the phospholipids and the structure of the protein which enters from one side. As shown in Table 7-10 there is a clear tendency in most reconstituted systems to one or the other orientation pattern. A random distribution, which would seem a priori the most probable, is the exception rather than the rule. In cases where phosphatidylcholine and phosphatidylethanolamine are required, a case can be made for a role of an asymmetric distribution of phospholipids as a contributing factor in the orientation of the protein. But in the case of the bacteriorhodopsin pump which operates in vesicles made with a single synthetic phosphatidylcholine such as dimyristoyl phosphatidylcholine, other factors must play a primary role such as the three-dimensional structure of the protein and the curvature of the functional membrane. Suggestive evidence for a role of the vesicle size and curvature of the membrane come from studies of particles obtained by sonic oscillation of *Halobacterium halobium.* In large vesicles the orientation of bacteriorhodopsin was similar to that in bacteria, pumping protons out, while the smaller vesicles were inside out pumping protons in (Kanner and Racker, 1975). Reconstitutions that permit a wide variation of experimental conditions should serve as a powerful tool in the analysis of factors involved in vectorial assembly.

Phospholipid Requirements

In collaboration with Dr. H. G. Khorana, we have started an investigation of the role of the phospholipid chemistry in reconstitutions. The first example tested (Knowles *et al.,* 1975) was

TABLE 7-11 **Acetyl Phosphatidylethanolamine in Reconstituted Systems**[a]

	nmoles (ng ions)/min/mg protein		
System	Phosphatidyl-ethanolamine	Acetyl phosphatidyl-ethanolamine	Acetyl phosphatidyl-ethanolamine + stearylamine
Proton pump	2.360	2.400	–
Calcium pump	114	0	109
P_i–ATP exchange	133[b]	0[b]	112[b]

[a]Experimental conditions were as described by Knowles *et al.* (1975).
[b]Phosphatidylcholine was also present during reconstitution.

acetyl phosphatidylethanolamine. This phospholipid readily formed vesicles and in reconstitutions with bacteriorhodopsin yielded an active proton pump. But neither the Ca^{2+} pump nor the mitochondrial P_i–ATP exchange was active when the acetylated lipid was used instead of phosphatidylethanolamine. However, incorporation of a simple alkyl chain with an amino group, e.g., stearylamine, together with acetyl phosphatidylethanolamine, resulted in the formation of active vesicles with either the Ca^{2+}-ATPase or the mitochondrial oligomycin-sensitive ATPase. These experiments are summarized in Table 7-11 and suggest that the amino group is required for the proper assembly or function of these two ion pumps.

With most membrane systems thus far examined by the cholate dialysis procedure, reconstitution is optimal with a ratio of 4:1 of phosphatidylethanolamine to phosphatidylcholine. In the case of the bacteriorhodopsin proton pump, a single phospholipid yields an active proton pump. In the case of the Na^+ pump of the plasma membrane, phosphatidylethanolamine alone gave high rates and inclusion of phosphatidylcholine lowered the rates of Na^+ translocation (Table 7-9). Objections could be raised to such determinations of phospholipid specificity because the proteins used for reconstitutions contain residual phospholipids (about 30% of the protein content). Although this represents less

TABLE 7-12 **Reconstitution of Delipidated Ca^{2+}-ATPase**[a]

	Ca^{2+} translocation (ng ions/min/mg protein)
Crude soybean phospholipids	216
Mitochondrial phosphatidylcholine (PC)	0
Egg phosphatidylcholine	0
Mitochondrial phosphatidylethanolamine	183
+ mitochondrial PC (1:1)	337
+ egg PC (1:1)	448

[a]The total amount of phospholipids in all experiments was 12 μmoles with 150 μg of Ca^{2+}-ATPase. Reconstitution was performed by the cholate dialysis procedure with 0.2 *M* potassium oxalate (Racker, 1972a). Other experimental conditions were as described by Racker *et al.* (1975a).

than 2% of the total phospholipids in the formed vesicles, the objections are valid if the residual phosphilipids in the proteins govern the properties of the immediate vicinity of the pump. Recent studies with a delipidated preparation of the Ca^{2+}-ATPase are therefore informative on this point (Racker *et al.*, 1975a). We have previously observed that the Ca^{2+} pump can be reconstituted with phosphatidylethanolamine as the only phospholipid but not with phosphatidylcholine alone (Racker, 1972a). It was reported by Warren *et al.* (1974) that replacement of the mixture of natural phospholipids in the isolated Ca^{2+}-ATPase by synthetic dioleoyl phosphatidylcholine sufficed for the reconstitution of ATPase and Ca^{2+} transport activity. We were able to confirm the reconstitution of the ATPase activity by phosphatidylcholine with preparations of ATPase from which we had removed over 95% of the phospholipids. But these vesicles did not catalyze active transport unless some phosphatidylethanolamine was included as shown in Table 7-12. Differences in experimental procedure may account for these divergent observations and one should remember that the natural sarcoplasmic reticulum membrane contains phosphatidylcholine as the major phospholipid with phosphatidylethanolamine, phosphatidylserine, and several other lipids including plasmalogens as minor constituents. Moreover, the protein–phospholipid ratio is much greater in the natural membrane than in our reconstituted vesicles. We have earlier pointed out (Racker and Eytan, 1973) that the artificial liposomes require inclusion of an inner Ca^{2+} trap such as oxalate or phosphate and do not respond like the natural membrane vesicles to the addition of external oxalate. On increasing the ratio of protein to phospholipid during reconstitution, the Ca^{2+} transport activity of the liposomes was greatly reduced, but now addition of external oxalate elicited a marked stimulation of Ca^{2+} transport. Thus the requirement for large amounts of phosphatidylethanolamine in reconstituted liposomes might be related to the specific assembly of the components in this artificial system and is obviously not relevant to the operation of the natural membrane. Nevertheless, reconstitution experiments can be most instructive in the elucidation of the specific role of phospholipids, of the effect of such minor components such as the plasmalogens, and of the role of the ratio between proteins and phospholipids in the formation of an active pump.

Mechanism of Action of Ion Pumps

Channel versus Mobile Carrier Mechanism

In Lecture 3 I have mentioned various mechanisms of ion translocation. Ion pumps reconstituted with synthetic phospholipids of known transition temperature lend themselves to the study of their mode of action. An example is the bacteriorhodopsin proton pump which operates in vesicles with single phospholipids such as dimyristoyl phosphatidylcholine (Racker and Hinkle, 1974). As shown in Fig. 7-4 proton translocation is abolished by nigericin, which exchanges H^+ against K^+, provided the membrane is fluid enough for this mobile carrier to traverse from one side of the membrane to the other. As the temperature is lowered and the membrane becomes more rigid, the pump operates even in the presence of nigericin, indicating that rhodopsin operates by a channel mechanism. When gramicidin, which is a channel former, is used instead of nigericin, ion translocation is abolished at all temperatures. With dipalmitoyl

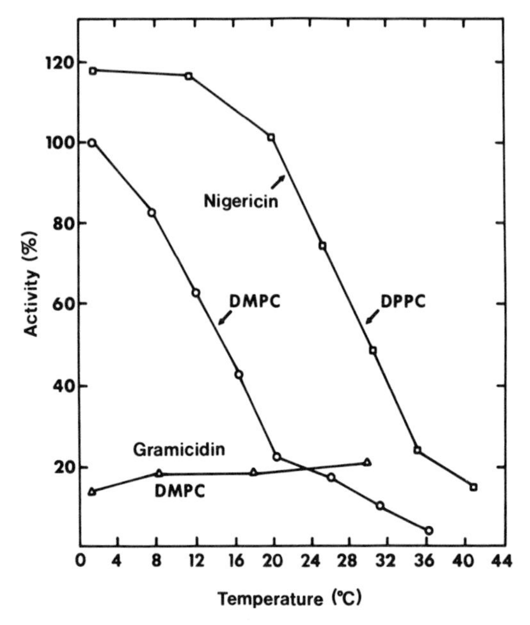

Fig. 7-4 Effect of temperature on the response of the bacteriorhodopsin proton pump to nigericin. Experimental conditions were as described by Racker and Hinkle (1974).

phosphatidylcholine which has a transition temperature about 20° higher than dimyristoyl phosphatidylcholine, the curve is shifted correspondingly to higher temperatures. Attempts to do similar experiments with other pumps that require phosphatidylethanolamine met with difficulties, partly because of the unavailability of phospholipids with suitable transition temperature that permit reconstitution of active vesicles, and partly because of the properties of the ATPase protein, which exhibits very high Q_{10} values for ATP hydrolysis and does not function well at very low temperatures.

The Rotating (Carrousel) Model

It has been proposed by a number of investigators (Deamer and Baskin, 1972; Makinose, 1973; Martonosi, 1973) that the phosphorylation of the Ca^{2+} pump ATPase is accompanied by a rotation of the protein resulting in the deposition of both Ca^{2+} and P_i on the inside of the sarcoplasmic vesicles. With reconstituted vesicles that are impermeable to P_i we could show decisively (Knowles and Racker, 1975b) that this mechanism is not operative. With ATP labeled with ^{32}P in the γ-position, an active Ca^{2+} translocation was not associated with the movement of $^{32}P_i$ into the vesicles. Although this experiment does not rule out every kind of rotating mechanism, it eliminates one particular formulation currently proposed for the Ca^{2+} pump.

Studies with Resolved Systems

The resolution of ATPases from membranes permits studies on the mode of interaction of the purified enzyme with cations, ATP, and P_i. Numerous reviews and books have been published on this subject (Tonomura, 1972; Nakao and Packer, 1973), and two basically different mechanisms have been proposed for the mechanism of action of ion translocating ATPases.

Mechanism I. The first formulation involves chemical enzyme intermediates and is backed by substantial experimental evidence obtained with the Na^+-K^+- and Ca^{2+}-ATPases. Evidence for the formation of a phosphoenzyme intermediate has been obtained from studies of an ADP–ATP exchange (Ebashi and Lipmann, 1962; Hasselbach and Makinose, 1962; Fahn *et al.,* 1966). When either the Ca^{2+}-ATPase or the Na^+-K^+-ATPase is incubated with ATP together with appropriate cations and then denatured, a phosphoenzyme can be detected with properties characteristic for an acyl group. An aspartyl residue of the protein has been identified as the bearer of the phosphoryl group (Bastide *et al.,* 1973; Degani and Boyer, 1973). Of particular interest is the formation of phosphoenzyme from $^{32}P_i$ in the presence of Mg^{2+} (Albers *et al.,* 1968; Lindenmayer *et al.,* 1968; Sen *et al.,* 1969; Masuda and de Meis, 1973) observed with membranous preparations of the Na^+-K^+- and Ca^{2+}-ATPase. Taniguchi and Post (1975) have obtained evidence for the formation of net ATP from

TABLE 7-13 **Formation of ATP by Ca^{2+}-ATPase and Its Utilization by Hexokinase**[a]

Treatment of product	Organic ^{32}P formed	
	− hexokinase	+ hexokinase
Exp. 1		
None	0.54[b]	0.49
Boiled 7 min in 1 *N* HCl	< 0.05	0.40
Exp. 2		
None	0.71[b]	0.92
Boiled 7 min in 1 *N* HCl	0.05	0.89

[a]Experimental conditions were as described by Knowles and Racker (1975a).

[b]Phosphoenzyme production determined in separate samples was 0.94 nmole/mg of protein in experiment 1 and 1.0 in experiment 2.

phosphoenzyme of kidney microsomal preparations on exposure to high concentrations of Na^+. It should be emphasized that these experiments are fundamentally different from those of previous investigators (Glynn and Lew, 1969; Makinose and Hasselbach, 1971; Panet and Selinger, 1972) who have demonstrated ATP formation by reversal of a functional ion pump. Not only has no one reported ion uptake with microsomal preparations of the Na^+-K^+-ATPase, but ATP formation takes place in an obligatory two-step reaction separated in time and is stoichiometric with phosphoenzyme. Moreover, ATP formation is associated with the addition of ions rather than with the dissipation of an ion gradient.

We have performed similar experiments with the Ca^{2+}-ATPase (Knowles and Racker, 1975a) with the following refinements. Phosphoenzyme was generated with a purified enzyme preparation that yields one major band on acrylamide gels (MacLennan, 1970). Phosphoenzyme was formed with $^{32}P_i$ in the absence of Ca^{2+}, which in fact inhibits the reaction. The presence of 3% Tween, which opens phospholipid vesicles, did not inhibit phosphoenzyme formation. In the second step ADP and Ca^{2+} were added simultaneously and [^{32}P]ATP was identified by several different methods. As shown in Table 7-13 the ATP was not formed by an exchange reaction, since it took place in the presence of a large excess of hexokinase which prevented a steady-state level of ATP by transfering the γ-phosphate to glucose. The yield of ATP which was formed was usually 50–80% of the phosphoenzyme and was free in solution, not bound to the enzyme.

These findings raise some serious thermodynamic questions. Where does the energy for ATP formation come from? The formation of phosphoenzyme from P_i has not been too disturbing since there is no direct information on the free energy of hydrolysis of the phosphoenzymes (E_2–P) formed from P_i. Calculations from equilibrium values (R. L. Post, personal communication) indicate that it is a "low energy phosphate," probably because the acyl phosphate is buried in a hydrophobic pocket of the protein and is not available for transfer to ADP. This is in contrast to E_1–P, the phosphoenzyme formed from ATP, which has been shown to transfer readily its phosphoryl group back to ADP (Post *et al.*, 1973). But a critical thermodynamic question arises when we demonstrate the formation of free ATP from P_i and ADP. After exploring other possibilities we have concluded that the binding energy of divalent cations to the protein must be the source of energy for ATP formation (Knowles and Racker, 1975a). The following scheme (Fig. 7-5, top line) is an outline of our present working hypothesis. In the first step Mg^{2+} interacts with the protein which undergoes a conformational change. Direct evidence for such a major conformational change at this step has been obtained in our laboratory by Dr. Y. Kuriki. Perhaps a thiol ester is formed during this step or activation of the carboxyl group of the aspartate residue takes place without covalent bonding. In the next step phosphoenzyme is formed with the phosphoryl group in a hydrophobic pocket, not available to either hydroxylamine or to ADP. On addition of Ca^{2+} a second conformational change in the enzyme takes place that exposes the phosphoryl group to the aqueous medium and renders it

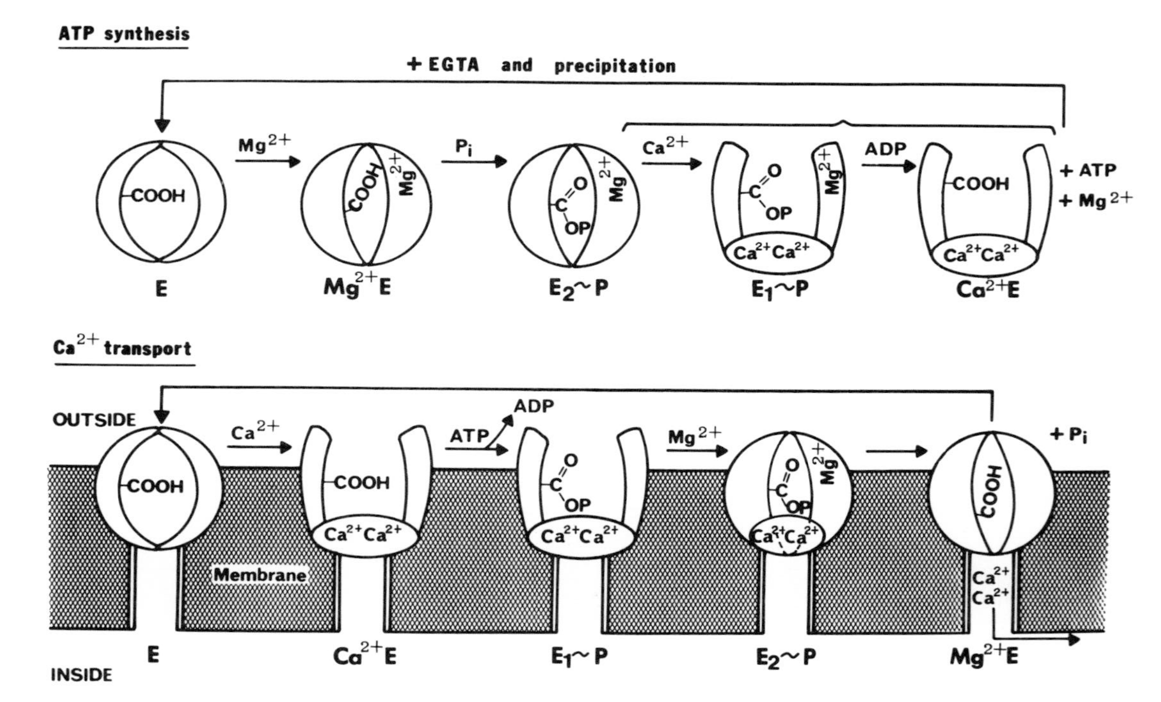

Fig. 7-5 Mechanism of ATP formation and Ca^{2+} translocation by Ca^{2+}-ATPase from sarcoplasmic reticulum. The channel attached to the ATPase in the bottom line is the proteolipid.

"high energy" and accessible to transfer to ADP. An extension of these considerations had led us to the formulation of Ca^{2+} translocation shown in Fig. 7-5 (bottom line). The interaction of Ca^{2+} and ATP gives rise to the formation of E_1–P with the phosphoryl group available to the water phase and with a firmly bound Ca^{2+}. The conformational steps that follow result in the release of the Ca^{2+} from the ATPase protein to the attached proteolipid "channel." Thus the ATP-driven reaction allows for Ca^{2+} accumulation within the channel and a release of Ca^{2+} down the gradient to the other side of the membrane. Our current experiments center around this formulation.

Mechanism II. In spite of the excellent chemical work performed on the phosphoenzymes of the Na^+-K^+- and Ca^{2+}-ATPases, there is no direct evidence available that they are intermediates either during ATP hydrolysis or ATP formation. Thus far, no one has succeeded in trapping a phosphoenzyme intermediate during catalysis without first inactivating the enzyme. An alternative interpretation proposes that the phosphoenzyme is on a side pathway or even an artifact caused during the denaturation or proteolytic digestion of the enzyme which precedes the demonstration of its existence.

Mitchell (1974) has proposed an alternative, chemiosmotic mechanism for ion translocating ATPases mainly based on studies of the mitochondrial proton pump. He proposes that the hydrophobic F_0 component of the pump (see Fig. 3-4) represents the oligomycin-sensitive proton conducting pathway through the membrane. F_1 is attached to F_0 in such a manner that its active center is exposed to the proton channel of F_0. Thus during hydrolysis of ATP the protons are released into the channel while ADP and P_i are translocated back via F_1 as monoanions to the water phase where they are protonated. In the reverse reaction during oxidative phosphorylation protons are picked up from the channel by phosphate at the active center of F_1. A concerted attack by ADP results in ATP formation. Mitchell has applied analogous formulations to the Na^+ pump and to other ion translocation systems.

The chemiosmotic model does not necessarily exclude a phosphorylated intermediate. Thus, during group translocation via the ATPase, a nucleophilic attack of the activated phosphate by an amino acid carboxyl group instead of by ADP can be readily visualized. In fact, such a mechanism might be used to control the directionality of the pump when ion transport rather than ATP generation is the primary physiological function, as is the case in the Na^+ and Ca^{2+} pumps.

The question whether mechanism I or II is operative in oxidative phosphorylation is the subject of an active controversy. Slater (1974) and Boyer (1974) have proposed a third mechanism which centers on the presence of tightly bound ATP in F_1 and CF_1. This firmly bound ATP would take the place of the phosphoenzyme intermediate in mechanism I. According to their formulations the release of the tightly bound nucleotide is the energy-requiring step in the formation of ATP from P_i and ADP. This release is proposed to be caused by a conformational change in the protein. There is no evidence for firmly bound ATP either in the Ca^{2+}-ATPase or in the Na^+-K^+-ATPase.

We are therefore confronted in the specific case of the proton pump of mitochondria with three alternative mechanisms illustrated in Fig. 7-6.

The first mechanism, which we can call the Mitchell mechanism, invokes a proton flux via F_0 and a direct interaction between phosphate and the proton flux with a concerted attack from the back of ADP as outlined above.

The other two mechanisms are indirect, requiring that the proton flux gives rise to a conformational change in the protein. In the second mechanism, which we might call the Boyer–Slater mechanism, ATP release is the result of the conformational change. In the third mechanism the conformational change gives rise to chemical changes at the active site and allows the interaction of P_i with the aspartyl residue to form a protein-bound acyl phosphate intermediate.

What are the arguments in favor of these various formulations? The Mitchell mechanism is the most difficult to prove, since it takes place by a concerted reaction involving a proton flux. Yet it

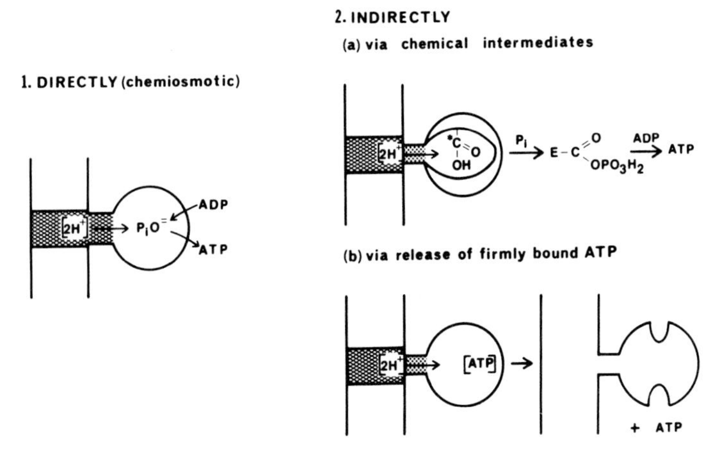

Fig. 7-6 Three mechanisms of ATP formation. In 2a the asterisk indicates an activation of the carboxyl group by a conformational change of the protein. In 2b the conformational change results in the release of ATP from the protein.

is of considerable interest that Boguslavsky *et al.* (1975) observed, with F_1 placed between an octane–water interface, a surface potential that is dependent on the presence in the octane phase of a hydrophobic proton acceptor, such as dinitrophenol. Another feature that follows from Mitchell's formulation is that the stalk of F_1 must be a proton channel connecting F_0 with the active site of the ATPase. This possibility can be approached experimentally.

The Boyer–Slater mechanism is attractive because it is consistent with the remarkable fact that several ATPases (from mitochondria, chloroplasts, and *E. coli*), associated with electron transport-linked phosphorylation, contain firmly bound ATP. Moreover, during energization of the membrane, e.g., by light in chloroplasts (Harris and Slater, 1975), the terminal phosphoryl group of ATP turns over. However, more kinetic data are required to ascertain that this turnover is rapid enough to be compatible with the requirements for an intermediate of photophosphorylation.

The phosphoenzyme intermediate hypothesis has no experimental support as far as the mitochondrial proton pump is concerned and is sustained merely by analogy with the Ca^{2+} and Na^{+}-K^{+} pump. There is considerable evidence for a phosphoenzyme intermediate in these two examples as pointed out earlier.

At the time of this writing I believe that it is not possible to decide which of the three mechanisms outlined in Fig. 7-6 is operative in the mitochondrial proton pump. We are at an impasse here that will require imaginative experimentation.

Control of Pumps

Reconstitutions lend themselves particularly for the study of controls in ion translocation. In the case of the Ca^{2+} pump we have observed marked differences in the efficiency of pumping as measured by the ratio of Ca^{2+} translocation over ATP hydrolysis. How is the control of the ATPase activity achieved in natural membranes to yield the observed high efficiency ratios? In sarcoplasmic reticulum fragments the Ca^{2+}/ATP ratio is 2. In our earlier reconstitution experiment the efficiency was low and the composition of the phospholipid and method of reconstitution had pronounced influences on the Ca^{2+}/ATP ratio (Racker and Eytan, 1973). The most intriguing influence is that of the proteolipid which profoundly regulates the efficiency of operation of the Ca^{2+} pump (Racker and Eytan, 1975).

Since control mechanisms in bioenergetics are the subject of the next lecture, we shall continue this discussion in connection with the metabolism of tumor cells.

Lecture 8

Control of Energy Metabolism

> Even if you are on the right track, you'll get run over if you just sit there.
>
> **Will Rogers**

Oxidation Control

In most mammalian cells that have been studied, the potential for energy production is much greater than energy utilization. The amount of ATP that could be generated by the available enzymes of the glycolytic and mitochondrial pathways is in vast excess of that needed for biosynthetic and work processes. It follows from this fact that a cell that wants to live economically must be able to control the consumption of energy-yielding substrates.

The basic principle of the control that governs both glycolysis and oxidation is brilliant and simple: *ATP is generated only when it is needed.* If only American industry would adopt this principle our economy would be sounder and I would be spared receiving electric back-scratchers as Christmas presents.

Oxidation Control in Glycolysis

Control of ATP generation during glycolysis has been extensively studied and is reasonably well understood (cf. Racker, 1965). Glyceraldehyde 3-phosphate is oxidized only when P_i and

ADP are available. As mentioned in the first lecture, the oxidation of the aldehyde substrates proceeds via a thiol ester–enzyme intermediate which is specifically cleaved by P_i. The product 1,3-diphosphoglycerate is removed by a phosphotransfer enzyme to ADP. The product of this reaction is 3-phosphoglycerate which also requires ADP for transformation to pyruvate. The latter serves as an acceptor for DPNH which is an allosteric inhibitor of glyceraldehyde-3-phosphate dehydrogenase. We shall return to the control of glycolysis when we discuss the Pasteur effect.

Oxidation (Respiratory) Control in Mitochondria

In the case of electron transport linked phosphorylation, according to Mitchell (1966), oxidation (or respiratory) control is achieved by the formation of a pH gradient and a membrane potential. This formulation is supported by the experiments on reconstituted systems which we described earlier; respiratory control is abolished only when compounds like nigericin and valinomycin, which collapse the ΔpH and the membrane potential, respectively, are used together. In mitochondria during operation of the Krebs cycle the proton motive force is used for ATP formation, thus allowing respiration to proceed as long as ADP and P_i are available. If the membrane is damaged and becomes leaky to protons, the proton motive force cannot be maintained and oxidation control is lost. We say such mitochondria are "loosely coupled." They respire in the absence of P_i and ADP, and the energy of oxidation is dissipated in heat.

The importance of respiratory control becomes immediately apparent when it ceases to function. Luft *et al.* (1962) have reported a case history of a loosely coupled woman or perhaps I should say of a woman who had loosely coupled mitochondria in her skeletal muscles. This unfortunate lady was incapable of proper function; her muscles worked with such inefficiency that she was severely incapacitated and bedridden. In some similar cases the problem of heat production by uncontrolled oxidation was so severe that continuous cooling of the patient was required.

Another problem of loose coupling was introduced by the

medical profession. During World War I it was noticed that personnel handling dinitrophenol derivatives in the course of the manufacture of explosives exhibited toxic symptoms, including weight loss. Actually dinitrophenols were used for coloring food (to indicate a high content of eggs!) at the end of the last century until the discovery of their toxicity forced the discontinuation of this fraudulent practice. At Stanford University systematic studies revealed the relationship between increased respiration, heat production, and weight loss caused by dinitrophenol. In spite of warnings from the Stanford scientists some enterprising physicians started to administer dinitrophenol to obese patients without proper precautions. The results were striking. Unfortunately in some cases the treatment eliminated not only the fat but also the patients, and several fatalities were reported in the *Journal of the American Medical Association* in 1929. This discouraged physicians for a while until they discovered that high doses of thyroxine resulted in weight loss. This was used for some time in the early 1930's until some more obituaries appeared. Now the physicians have become more restrained and use "mild"-miracle diets to cure obesity. Sometimes the diets are not so mild either.

One of these miracle cures was published in a best-seller by H. Taller entitled "Calories Don't Count." I remember one evening about 12 years ago when one of my wife's cousins had dinner in our home. He proclaimed that he followed Taller's recommendation and that he could now eat unlimited quantities of food. His demonstration that night was impressive. Since at that time he was an assistant professor at Princeton and I was not even at Cornell, I was in no position to argue with him, let alone convince him that Taller's thesis of "calories don't count" was sheer nonsense. I did not sleep too well that night, and the next morning I announced to my wife that I had decided to write a review of Taller's book. Her objections that I hadn't read the book didn't bother me, and I submitted a short article to the *American Journal of Medicine* which was published as an editorial under the title "Calories Don't Count, if You Don't Use Them". I explained that the essential ingredient of Taller's diet was a high content of food with unsaturated fatty acids which, like dinitrophenol and thyroxine, are uncouplers of oxidative phosphorylations. While in some cases these uncouplers may be effective, their clinical control is very difficult because of the narrow range between tolerated and toxic doses.

The fact that fatty acids are both natural intermediates of fat metabolism and uncouplers of oxidative phosphorylation raises the interesting problem of their potential physiological role. It is conceivable that the large individual variation in tolerating large intake of food without weight gain may be related to differences in the relative capacity to produce and remove free fatty acids. It is an interesting problem that needs more investigation.

There are two examples pointing to an active role of fatty acids as natural uncouplers. The brown fat of animals that are hibernating or exposed to cold for extended periods contains large amounts of mitochondria (which are responsible for the color). Earlier studies showed that brown fat mitochondria did not catalyze oxidative phosphorylation (Smith *et al.,* 1966). This was later traced to the effect of fatty acids which could be counteracted by addition of bovine serum albumin (Guillory and Racker, 1968; Rafael *et al.,* 1968). It seems likely that these animals develop brown fat and uncouple oxidative phosphorylation for the purpose of producing heat. Another example is the involution of the lactating mammary gland. It appears that the signal for the regression of this organ is given by fatty acids which throttle the energy supply (Nelson *et al.,* 1962).

The Pasteur Effect

I want to discuss in considerable detail the high aerobic glycolysis of tumor cells. Before we can analyze the loss of control of glucose utilization in tumor cells, we have to discuss the control mechanisms that operate in normal cells. Pasteur discovered (1861) that, in the presence of air, yeast cells consumed less glucose per unit weight than under anaerobic conditions. Numerous hypotheses have been proposed to explain the Pasteur effect, and we can learn interesting lessons from a study of the history of the investigations that focused on this phenomenon (Racker, 1974). There are few subjects in biochemistry that have been more confusing to students and more seductive to wild speculations. One of the reasons why early studies went astray can be traced to methodology. If investigators had used the approach of Pasteur measuring glucose utilization, we would have been spared a lot of confusion. Instead, seduced by Warburg's elegant and convenient manometric methods for the measurement of acid production, everyone determined the appearance of lactic acid rather than the disappearance of glucose. Thus, most of the thinking was expended on explanations for the diminished lactate production under aerobic conditions.

The Competition Mechanism

Historically the most important hypothesis of the Pasteur effect was proposed independently by Johnson (1941) in the United States and by Lynen (1941) in Germany. Both emphasized a key role of inorganic phosphate in regulating glycolysis. The idea was that glycolysis and oxidative phosphorylation compete for inorganic phosphate; in the presence of oxygen mitochondrial metabolism prevails and thereby limits the availability of P_i for glycolysis. This was a simple and brilliant idea that quickly became popular. The only thing wrong with it was that it only explained an inhibition of lactate formation which was dependent on P_i, but it did not explain the original Pasteur effect on the inhibition of glucose phosphorylation which is independent of P_i and requires only ATP, a product of oxidative phosphorylation.

Allosteric Control by ATP and Glucose 6-Phosphate

It took about ten years for this thought to penetrate and explanations proposed at a Ciba symposium (Chance, 1959; Lynen *et al.,* 1959; Racker and Wu, 1959) centered around a mitochondrial compartmentation of ATP which was first considered by Lynen and Königsberger (1951). However, these formulations were abandoned when evidence for allosteric control of phosphorylation started to accumulate. Experiments by Lonberg-Holm (1959) and Lynen *et al.* (1959) with intact cells and by Lardy and Parks (1956), Passonneau and Lowry (1962), Rose *et al.* (1964), and Uyeda and Racker (1965a, b) with isolated enzymes led to a rational concept of a coordinated control of glucose phosphorylation. Hexokinase is inhibited by glucose 6-phosphate, and phosphofructokinase is inhibited by high concentrations of ATP. These two enzymes are inhibited at concentrations of the allosteric effectors which are physiological, and these two intermediates therefore influence the rate of glucose utilization. The Pasteur effect can thus be formulated as the result of the sequential and cascading events that take place during oxidative phosphorylation: (a) depletion of inorganic phosphate and ADP required for glycolysis and increase in the ATP level, (b) inhibition of phosphofructokinase by ATP resulting in accumulation of glucose 6-phosphate, and (c) inhibition of hexokinase by glucose 6-phosphate. I would like to illustrate some of these points by experimental results. We have reconstituted glycolysis by putting together the individual enzymes of the pathway (Gatt and Racker, 1959a,b). In the presence of catalytic amounts of P_i and ADP, the rate of lactate production in this system was greatly stimulated by addition of an ATPase (Fig. 8-1). When too much ATPase activity was added, the rate of glycolysis was depressed. For continuous glycolysis the rate of hydrolysis and regeneration of ATP had to be equal in order to achieve steady-state concentrations of P_i, ADP, and ATP.

The Role of P_i

We have also shown that in intact ascites tumor cells the major rate-limiting factor in glycolysis is inorganic phosphate (Wu and Racker, 1959a,b). Any condition that increased P_i, e.g., high external P_i or addition of an uncoupler of oxidative phosphorylation, greatly increased the rate of lactate production. The mecha-

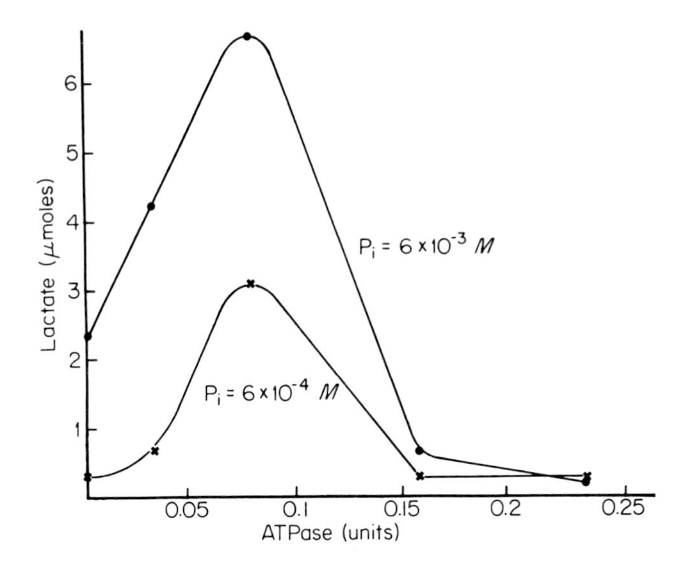

Fig. 8-1 Effect of ATPase addition on a reconstituted system of glycolysis. Experimental conditions were as described by Gatt and Racker (1959b). Lactate formation was measured at low and high P_i concentration in the presence of increasing amounts of an ATPase (isolated from potatoes).

nism of this stimulation by phosphate is complex because not only was lactate production stimulated but glucose utilization as well. Later it was shown that P_i is also an allosteric effector stimulating phosphofructokinase which is inhibited by excess ATP (Passonneau and Lowry, 1962; Uyeda and Racker, 1965a). P_i also reverses the inhibition of hexokinase by glucose 6-phosphate (Rose *et al.,* 1964; Uyeda and Racker, 1965a).

However, this simple coordinated control mechanism is not a universal feature and I have pointed out repeatedly (Racker, 1954, 1974) that there are probably a large number of "Pasteur inhibitors" that participate in the control of hexose utilization. Our picture of the mechanism of the Pasteur effect as outlined above is probably basically correct but greater oversimplified.

Secondary Allosteric Effectors

The complexities and diversities of the mechanism that participate in the operation of the Pasteur effect are illustrated by two striking examples. It is clear that in yeast, the cells that were first shown to have a Pasteur effect, the above widely accepted mechanism (Krebs, 1972) is not operative because yeast hexokinase is not inhibited by glucose 6-phosphate. Unpublished experiments from our laboratory suggest that a metabolic product of glucose 6-phosphate, perhaps mannose 6-phosphate, may be the allosteric effector of yeast hexokinase. Alternatively, an inhibition of glucose entry by glucose 6-phosphate may participate in the Pasteur effect (Sols, 1967). A more serious complication is the fact that in yeast cells as well as in certain mammalian tissues such as muscle, or in Ehrlich ascites tumors, analysis of the steady-state concentration of ATP did not reveal significant changes during transition from the anaerobic to the aerobic phase. Obviously, there are still missing links that need to be discovered. In the case of skeletal muscle, the ATP concentration is kept constant by regeneration from the phosphocreatine store. Indeed, on examination of the effect of phosphocreatine on phosphofructokinase we found an inhibition at levels of phosphocreatine that may be effective *in vivo* (Uyeda and Racker, 1965a). However, more direct evidence for a physiological role of phosphocreatine as a secondary allosteric effector is needed. There have been other reports of metabolites that inhibit phosphofructokinase, but in most instances the concentrations required for inhibition are so great compared to physiological concentrations that it is difficult to make a convincing case for their operation as Pasteur inhibitors. On the other hand, there is evidence for synergistic action among allosteric effectors which makes it a little easier to consider the significance of inhibitors such as NH_4^+, phosphoenolpyruvate, or citrate which affect the enzymes only at relatively high concentrations under the artificial conditions of the *in vitro* assay.

The scheme of the mechanisms of the Pasteur effect outlined in Fig. 8-2 therefore represents only a tentative scheme that still

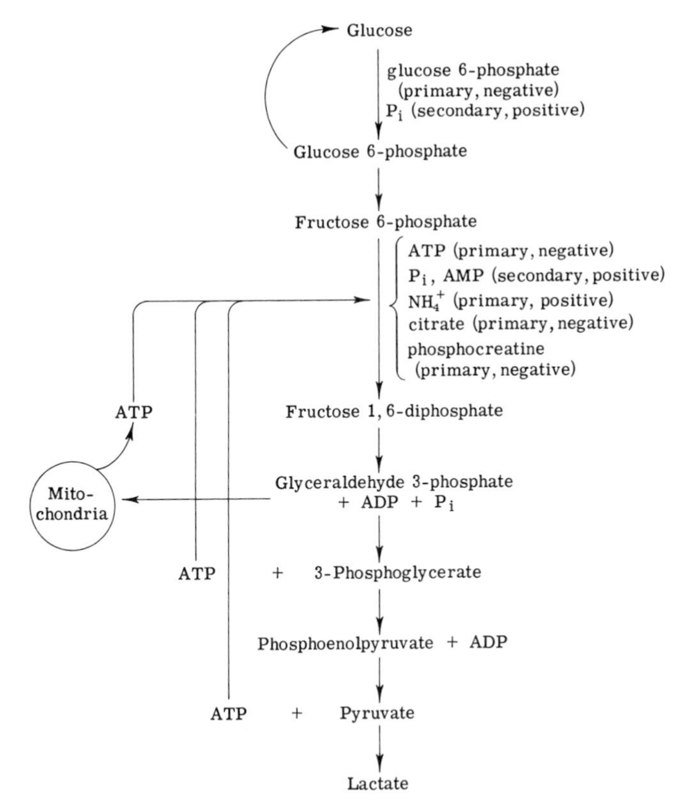

Fig. 8-2 Allosteric control of phosphofructokinase and hexokinase participating in the Pasteur effect.

requires amplification. In this scheme I refer to positive and negative effectors dependent on the stimulatory or inhibitory action on the enzyme. Primary effectors act directly on the enzyme; secondary effectors counteract a stimulation or inhibition of a primary effector.

The basic sequence is visualized as follows: The primary event is the competition of oxidative phosphorylation for ADP and P_i resulting in an inhibition of lactic acid formation. In various cells the consequence of this may differ depending on the ATPase activity, regeneration of ATP from phosphocreatine, uptake of NH_4^+ during α-ketoglutarate oxidation, etc. The overall result is either an increase in negative effectors or a decrease in positive effectors (e.g., P_i and NH_4^+) resulting in a lower activity of phosphofructokinase and glucose 6-phosphate accumulation. This in turn shuts off glucose utilization either by inhibiting hexokinase directly or via a metabolite. Alternatively, glucose transport may be inhibited. It is quite likely that in many instances there is a complex orchestration of these various effectors resulting in the final effective control of carbohydrate metabolism (cf. Racker, 1975).

High Aerobic Glycolysis in Tumor Cells (The Warburg Effect)

About fifty years ago Otto Warburg (1926) discovered that tumor cells convert more glucose to lactic acid under aerobic conditions than normal cells. Although Warburg's observations were soon confirmed in many other laboratories, the phenomenon itself, which I refer to as the Warburg effect, has been largely ignored. It may be instructive to analyze the reasons for this. Warburg is generally referred to as one of the great biochemists of this century. Yet his views on the two major problems that he was interested in, namely, cancer and photosynthesis, are widely held to be both naive and incorrect. I think we can find the reason for this in Warburg's personality. I have mentioned earlier his formulations of the mechanism of action of glyceraldehyde-3-phosphate dehydrogenase which were simple, brilliant, and wrong. In 1952 we published our work on the participation of a thiol ester during oxidation of glyceraldehyde 3-phosphate (Racker and Krimsky, 1952). Warburg *et al.* (1954) wrote an article on "Racker's Umweg" (meaning detour) in which he rejected the thiol ester mechanism as being too complicated, but presented no evidence against it. I remember my reaction when I first read his article. I was still (relatively) young and unknown and I confess that I felt a little flattered by the attention of the great biochemist, but of course also annoyed and shocked by his complete lack of objectivity. I seriously considered writing a rebuttal but soon thought it would be better to let time take its course. I have never regretted that I avoided a polemic with Warburg. Instead I became interested in his psychological makeup and discovered other examples of the arrogance of his mind. Some time ago, Warburg had suggested that a carboxyl group becomes attached to chlorophyll during CO_2 fixation in plants. When experiments in his own laboratory showed no $^{14}CO_2$ incorporation into chlorophyll, he concluded that the widely accepted structural formula of chlorophyll had to be in error!

In the case of the cancer problem, his rigid mind became an obstacle in the recognition of the importance of his discovery. He proposed that the reason for the high aerobic glycolysis is a damage to the respiratory chain resulting in a reversion of the cell to the more primitive fermentative metabolism similar to that of an anaerobic yeast. This emphasis on a damaged respiratory chain led to controversy (Weinhouse, 1956) because many tumor cells have respiration and coupled phosphorylation quite comparable to that of normal cells. Thus in the discussions of the Warburg effect the emphasis was placed on the wrong hypothesis and away from the experimental finding. Warburg's views were so discredited that in the recent National Cancer program high aerobic glycolysis of tumor cells could not find its way into the giant wheel of the "research strategy hierarchy."

I have made an effort to rescue the "Warburg effect" by formulating an alternative and broader hypothesis (Racker, 1972a). Since we knew that the rate-limiting factor for lactate formation is the supply of P_i and ADP, I proposed that the high aerobic glycolysis of tumor cells is caused by the activation of an ATPase which may be in the mitochondria or in the plasma membrane or may even be in a virus. Since there appears to be a rough correlation in experimental tumors between the rate of glycolysis and malignancy, it is conceivable that the high aerobic glycolysis gives rise to an altered metabolic state in the cell that releases the constraints on growth that prevail in normal cells. The most obvious parameter is the intracellular pH which is lowered during glycolysis and which is known to influence many enzyme catalyzed reactions as well as their allosteric controls. But other parameters should be considered as well, such as the ratio of adenine nucleotides or the intracellular content of free P_i.

ATPases in Tumor Cells

I mentioned earlier that ATPase is a glycolytic enzyme; without regeneration of ADP and P_i the formation of lactate from glucose comes to a standstill. I should emphasize that I refer to an "ATPase" in the broadest sense, including dissipation of ATP in work processes such as muscular contraction and ion translocation. A list of various ATPases that are potential participants in glycolysis are shown in Table 8-1. Although I have included in this list biosynthesis of proteins, nucleic acids, etc., I doubt that they make a notable contribution to ATP hydrolysis in tumor cells where these processes are usually very slow compared to the rate of glycolysis. Even in fairly rapidly growing tumor cells the energy expended on biosynthetic processes is small compared to that used, for example, for the transport of ions. Moreover, we observed only small interferences with glycolysis in the presence of inhibitors of protein or nucleic acid synthesis.

TABLE 8-1 **Potential ATPases Participating in Glycolysis**

ATPase	Activation
Mitochondrial ATPase	Dissociation of inhibitor, uncouplers (fatty acids)
Na^+-K^+-ATPase (plasma membrane)	Loss of control (proteolipid?)
Ca^{2+}-ATPase (sarcoplasmic reticulum)	Loss of control (proteolipid)
Virus associated ATPase	Infection
Biosynthetic processes	Hormonal, loss of control
Futile cycles	Hormonal, protein alteration
Other ATPases (lysosomal,Ca^{2+}-ATPase) of plasma membrane, etc.	Unknown
Other energy-dependent reactions (transport, reductions, etc.)	Unknown

TABLE 8-2 **Effect of Na^+, K^+, and Ouabain on Glycolysis of Ascites Tumor Cells**[a]

Additions	μmoles lactate/30 min/mg	
	− dinitrophenol	+ dinitrophenol
Ascites tumor cells in buffer	0.36	1.24
+ rutamycin	0.43	0.41
+ ouabain	0.12	1.03
− K^+	0.16	−
− Na^+	0.15	−
− K^+, Na^+	0.04	−

[a]Experimental conditions were as described by Scholnick *et al.* (1973).

The question which we attempted to answer first is which of the various ATPases actually sustain the high aerobic glycolysis of tumor cells. During the past years we have examined several tumor lines and it has become quite apparent that different ATPases can participate in the high aerobic glycolysis. To illustrate our approach to this problem I shall describe our studies on the Ehrlich ascites cells, the first tumor we analyzed. We selected two inhibitors, rutamycin which inhibits mitochondrial ATPase and ouabain which inhibits the Na^+-K^+-ATPase of the plasma membrane. As shown in Table 8-2 rutamycin did not inhibit lactate formation, but ouabain did. We were first worried that such a high concentration of ouabain was required for inhibition, but measurements of Rb^+ uptake convinced us of the close relationship between inhibition of pump activity and of aerobic glycolysis (Scholnick *et al.*, 1973). Moreover, when the mitochondrial ATPase was activated by dinitrophenol, the stimulated glycolysis was sensitive to rutamycin but insensitive to ouabain. When we arrive at conclusions based on the use of inhibitors such as ouabain, we like to obtain additional evidence by some other means. As shown in Table 8-2 deletion of either K^+ or Na^+ from the medium depressed the rate of glycolysis. When both ions were

TABLE 8-3 **Control of Glycolysis in Various Cell Lines**[a]

Additions	L-1210	P-388	3T3	PY-3T3	BHK	PY-BHK
None	0.40	0.40	0.41	0.86	0.6	0.7
Ouabain	0.24	0.11	0.60	0.70	0.5	0.7
Dinitrophenol	0.68	1.65	0.69	1.40	1.5	2.0
Rutamycin	0.48	0.57	0.18	0.27	1.2	1.7

[a]Experimental conditions were as described by Suolinna *et al.* (1975).

deleted, glycolysis was virtually eliminated. Since dinitrophenol stimulated glycolysis in the presence of ouabain or in the absence of K^+ in the medium, it was clear that there was sufficient K^+ in the cell to sustain a high rate of glycolysis. The clear implication of these experiments is that the Na^+-K^+-ATPase of the plasma membrane is responsible for generating the ADP and P_i required for glycolysis.

A survey of several cell lines shown in Table 8-3 established that cells differ greatly in the response to various inhibitors. Glycolysis in 3T3 cells or polyoma transformed 3T3 cells was inhibited by rutamycin suggesting that mitochondrial ATPase sustained glycolysis. Ouabain had little or no effect on these cells. In baby hamster kidney cells, none of the conventional ATPase inhibitors had any effect on glycolysis. We do not know as yet which ATPase is responsible for the high glycolysis in these cells, but we do know that it is very sensitive to quercetin, an ATPase inhibitor that we shall discuss shortly.

At this stage of our research we asked ourselves whether we could devise experiments to test the hypothesis that a high aerobic glycolysis contributes to the malignancy of tumor cells. We felt that compounds that inhibit glycolysis directly would be too toxic since many organs, particularly the brain, are very dependent on glucose as a major energy source. We thought we should inhibit glycolysis in tumor cells by repairing specifically the defect that is responsible for the high rate of glycolysis. The first question we had to answer is whether the increased ATPase activity associated with the Na^+-K^+ pump or the mitochondria was the result of a more active or of a less efficient ion translocation. Do tumor cells translocate 3 Na^+ and 2 K^+ ions for each ATP that is hydrolyzed or do they have ratios that are different from those reported for red blood cells or nervous tissue (cf. Post *et al.*, 1973)? It is difficult to measure these ratios accurately in intact cells, but an answer was obtained by the following indirect method which was used by Whittam and Ager (1965). When ATP is utilized by the Na^+-K^+ pump, one lactate can be generated by glycolysis for each ADP released. If we measure Rb^+ uptake and lactate formation, the ratio of Rb^+/lactate can be used as an index of the efficiency of the pump provided (a) ADP and P_i are the rate-limiting factors for glycolysis, (b) the Na^+-K^+-ATPase is the major supplier of ADP, and (c) utilization of ADP by other systems is relatively minor; if it is appreciable the efficiency of the pump will appear higher than it actually is. Our studies showed that in Ehrlich ascites cells the ratio was much below 2, usually between 0.5 and 1 (Suolinna *et al.*, 1974). This is likely to be an upper estimate of the efficiency since some of the ADP and P_i generated by the pump was used for mitochondrial ATP production, thereby reducing the yield of lactate. We concluded that the aerobic glycolysis can be largely accounted for by a low efficiency of ion translocation rather than by a major increase in Na^+-K^+ pump activity.

Repair of Ion Pumps in Tumor Cells

We therefore searched for a compound that would inhibit the extra ATPase activity of the Na^+-K^+ pump but not interfere with ion translocation. We wanted a compound that acted in a manner similar to the mitochondrial ATPase inhibitor, but which had a low molecular weight and permeability properties that would allow its assay in intact cells. We felt that an inhibitor of the Na^+-K^+-ATPase like ouabain was not suitable because it also inhibits ion translocation of the normal cells and therefore is quite toxic. We found a group of compounds that met our specifications among the bioflavonoids which are natural plant

TABLE 8-4 **Inhibition of Mitochondrial ATPase by Bioflavonoids**[a]

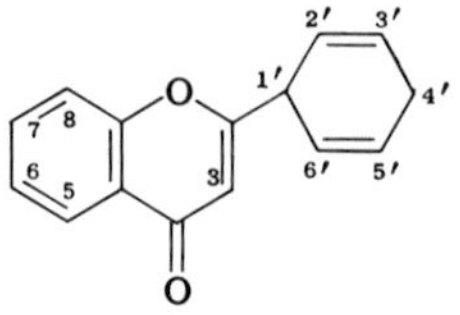

Flavone

		ATPase 50% inhibition (μg)
Myricetin	3,3′,4′,5,5′,7-OH	6
Quercetin	3,3′,4′,5,7-OH	8
Fisetin	3,3′,4′,7-OH	11
Morin	2′,3,4′,5,7-OH	40
Rutin	quercetin-3-rhamnoglucoside	not inhibitory

[a]Experimental conditions were as described by Suolinna *et al.* (1974).

products and ingredients of common foods. Some representative bioflavonoids and their effect on mitochondrial ATPase activity are listed in Table 8-4. Quercetin is the bioflavonoid that we have tested most extensively mainly because it is commercially readily available and quite active. Surprisingly, it acts both on mitochondrial and Na^+-K^+-ATPase in a similar manner by inhibiting ATP hydrolysis at low concentrations without affecting oxidative phosphorylation or ion translocation. As shown in Table 8-5, oxidative phosphorylation was not affected at quercetin concentrations below 20 μg/ml which inhibited ATPase activity quite markedly. It acted similarly to the natural mitochondrial ATPase inhibitor. Higher concentrations of quercetin also inhibited oxidative processes. In ascites tumor cells glycolysis was inhibited by quercetin (Table 8-6), but under appropriate conditions Rb^+ uptake was not inhibited. This resulted in Rb^+/lactate ratios often

TABLE 8-5 **Effect of Quercetin on Oxidative Phosphorylation**[a]

Additions	ATPase activity (μmoles P_i/10 min)	P:O
None	0.83	0.96
Quercetin 12 μg	0.31	1.04
20 μg	0.28	1.02
F_1 inhibitor 0.9 μg	0.54	1.04
2.1 μg	0.38	1.1

[a]Experimental conditions were as described by Lang and Racker (1974).

exceeding 2, which is an overestimate as pointed out above. In the presence of dinitrophenol which eliminated the mitochondrial competition for ADP, the Rb^+/lactate ratios were close to 2.

When we tested the effect of quercetin on the growth of tumor cells in tissue cultures, we ran into two complications. Both serum albumin and bicarbonate, constituents of the conventional growth medium, counteracted the effect of quercetin. We therefore used Hepes as the main buffer (Eagle, 1971) and reduced the amount of bicarbonate and serum in the culture medium. Polyomavirus transformed baby hamster kidney cells or leukemia 1210 cells failed to proliferate in the presence of 10 to

TABLE 8-6 **Effect of Quercetin on Rb+/Lactate Ratio**[a]

Additions	(μmoles/30 min/mg protein)		
	Lactate formation	Rb^+ uptake	Rb^+/lactate ratio
None	0.36	0.30	0.83
Quercetin (16 μg/ml)	0.12	0.36	3.00
Dinitrophenol	2.08	0.37	0.17
Dinitrophenol plus quercetin	0.19	0.37	1.95

[a]Experimental conditions were as described by Suolinna *et al.* (1975).

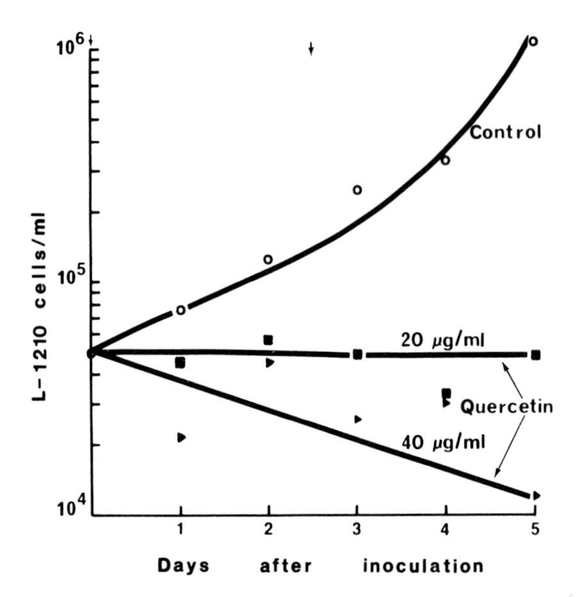

Fig. 8-3 Inhibition of growth of L-1210 cells by quercetin. Experimental conditions were as described by Suolinna *et al.* (1975).

20 μg of quercetin per milliliter of culture media (Fig. 8-3). So-called "normal" (untransformed) cells were inhibited also. Although we observed some quantitative differences in susceptibility, the sensitivities were too variable to draw any conclusions except that both "normal" and transformed cells were inhibited. There are serious questions in the minds of most investigators whether cells grown in plastic plates, living in a ghetto, as Leonard Warren puts it, are representative of "normal" cells.

We then examined a large number of bioflavonoids isolated from different plants and sent to us from all over the world in the hope of finding one that would affect the growth of tumor cells even in the presence of bicarbonate and serum albumin, since any candidate for an *in vivo* assay would have to be effective in their presence. Among over 100 compounds tested we found two active inhibitors of tumor glycolysis which remained effective in the presence of bicarbonate and were only partly counteracted in

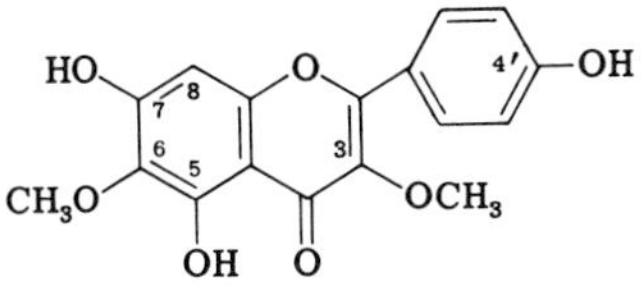

(a)

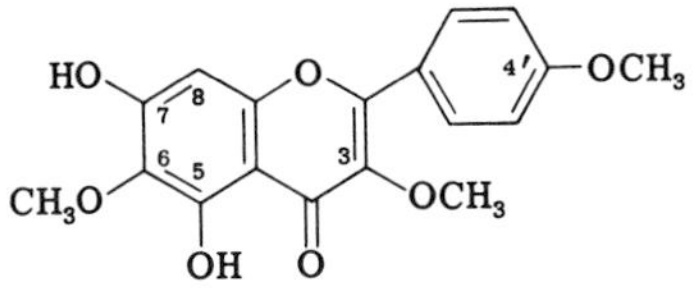

(b)

Fig. 8-4 Chemical structures of methoxyflavonoids effective as inhibitors of glycolysis in ascites tumor cells. (a) From Wollenweber and Lebreton (1971). (b) From Goudard and Chopin (1974).

the presence of serum. These compounds are methoxy derivatives of flavonoids. The natural product was sent to us by Dr. Wollenweber from Germany and a related synthetic sample was given to us by Dr. Chopin from France. The chemical structure of these compounds is shown in Fig. 8-4. When we tested the effect of the methoxy quercetins on mitochondrial and Na^+-K^+-ATPase, we found to our surprise that they inhibited neither! We first suspected that these compounds must be acting by some other mechanism such as inhibiting one of the glycolytic enzymes. To check on this point we used ascites tumor cells that had been treated with dextran sulfate. To explain this I have to digress.

In the course of our studies on the rate-limiting factors in tumor glycolysis, we became very interested in the rate of phosphate transport (Racker, 1965). We observed in 1963 that dextran sulfate changes the permeability properties of ascites tumor cells, and a recent more extensive analysis revealed that under controlled conditions the cells can be made permeable to phosphate and nucleotides in a reversible manner (Scholnick *et al.*, 1973). Glycolysis in dextran sulfate treated cells was very low, but could be restored to original or sometimes even higher levels

TABLE 8-7 **Effect of Dextran Sulfate on Glycolysis and Adenine Nucleotide Content of Ascites Tumor Cells**[a]

Additions	Lactate formation (μmoles/30 min/mg protein)	Intracellular nucleotides (nmoles/mg protein)		
		ATP	ADP	AMP
Ascites cells	0.46	23	6.8	1.5
+ AMP	0.48	33	7.0	0.9
Dextran sulfate treated cells	0.20	4.5	3.8	0.8
+ AMP	0.37	11	6.8	3.5

[a]Experimental conditions were as described by Scholnick *et al.* (1973).

if AMP as well as P_i was added to the assay mixture (Table 8-7). This is in line with the finding that the nucleotide level in these cells was low.

When we checked the viability of these cells, we found that they were as tumorigenic as control cells. An analysis of dextran-treated cells that were injected into the peritoneal cavity of mice showed that they quickly recovered their glycolytic activity and dextran sulfate was no longer attached (Table 8-8). In pursuing these observations, it was observed that a short incubation of dextran-treated cells with ascites fluid rapidly reversed the permeability changes as shown in Table 8-9 (McCoy *et al.,* 1976).

These reversibly damaged ascites cells have been most useful in the study of processes that depend on nucleotides or other small molecular ions that can be supplied in the medium. Thus, glycolysis was lost in dextran sulfate-treated cells, but was restored on addition of P_i and AMP. As shown in Table 8-9 ouabain or quercetin (or methoxyflavonoids) had no effect on glycolysis under these conditions, but sensitivity to the inhibitor was restored when the cells were repaired by incubation with ascites fluid. These experiments showed that these compounds did not inhibit the glycolytic pathway per se, but interfered with the

TABLE 8-8 ***In Vivo*** **Restoration of Glycolysis in Dextran Sulfate-Treated Ascites Tumor Cells**[a]

Time after inoculation of cells into mice (minutes)	Rate of lactate formation (μmoles/30 min/mg protein)		$[^3H]$Dextran sulfate bound/mg protein (μg)
	– AMP	+ 5 m*M* AMP	
0	0.09	0.30	0.46
30	0.29	0.30	0.20
90	0.34	0.38	0.04
Untreated cells	0.28	0.30	0

[a]Experimental conditions were as described by McCoy and Racker (1976).

generation of ADP and P_i. Since quercetin and the methoxy derivatives behaved similarly in these experiments, it was rather puzzling that quercetin was an inhibitor of isolated Na^+-K^+-ATPases from various sources while methoxyflavonoid was not. The solution to this problem came when we prepared an Na^+-K^+-ATPase from the ascites cells. The fresh membrane preparation was sensitive to methoxyflavonoid while aged prepara-

TABLE 8-9 **Repair of Dextran Sulfate-Treated Ascites Tumor Cells with Ascites Fluid**[a]

Addition	Lactate formation (μmoles/30 minutes/mg protein)		
	Untreated cells	Dextran sulfate-treated cells	Repaired cells
P_i buffer	0.24	0.07	0.21
+ AMP (5 m*M*)	0.26	0.31	0.27
+ quercetin	0.13	0.29[b]	0.15
+ ouabain	0.14	0.27[b]	0.11

[a]Experimental conditions were as described by McCoy and Racker (1976).

[b]Assayed in the presence of 5 m*M* AMP.

TABLE 8-10 **Effect of Quercetin on Protein Synthesis in Ascites Tumor Cells**[a]

Cell suspension	$[^{14}C]$Valine incorporation into proteins (cpm/30 min/mg protein)	
	− quercetin	+ quercetin
Ascites tumor cells	55,160	19,080
Dextran sulfate-treated cells	29,170	28,730

[a]Experimental conditions were as described by McCoy and Racker (1976).

tions were sensitive only to quercetin. This phenomenon is now under study.

At present we are also examining why quercetin inhibited the growth of cells in tissue cultures. As shown in Table 8-10, quercetin at rather low concentrations inhibited protein synthesis of ascites tumor cells in suspension. After treatment with dextran sulfate, little or no inhibition by quercetin was observed (McCoy and Racker, 1976). Our first suspicion that the control mechanism was linked to lactic acid formation resulting in a change in pH proved to be wrong. At least in dextran sulfate-treated cells we could not observe a stimulation of protein synthesis by lowering the pH to values operative in ascites tumor cells. The second possibility that a control may be exerted by P_i or by the ATP/ADP ratio seems at present more likely.

Before leaving the subject of the control of glycolysis by quercetin which acts as an artificial regulator of the pump, I would like to discuss briefly the natural control of ion pumps. In the case of the mitochondrial ion pump we have learned that a small molecular weight protein inhibits the hydrolytic activity of the ATPase without interfering with oxidative phosphorylation. How are the other pumps regulated and what changes in the membranes of tumor cells are responsible for a loss of control and for a low efficiency of pump operation?

It is remarkable that current journals in the area of bioenergetics publish dozens of papers on the mechanism of action of ion pumps and virtually none on their control. Yet, the problem is confronting us. Although we have no decisive answers, experiments on reconstituted ion pumps have yielded important clues. We have learned that the composition of the phospholipids and conditions of reconstitution profoundly affect the efficiency of the Ca^{2+} pump operation (Racker and Eytan, 1973). Moreover, as discussed earlier, the proteolipid markedly increased the efficiency of Ca^{2+} transport in reconstituted vesicles. We are just at the beginning of these investigations of pump control. We can only hope that the reconstituted systems are representative of events in the natural membrane in this respect and will yield meaningful information. This is a chance we take, but after all we have taken chances for almost a century when we studied enzymes that had been removed from their natural environment.

You can see that we are still in the middle of a research project that we started 20 years ago when we began to study glycolysis in ascites tumor cells (Racker, 1956). It has led us into various aspects of fundamental biochemistry, including development of enzyme assays, reconstitution of glycolysis, studies of allosteric controls, control and mechanism of pump action, and, now, control of protein synthesis. The story makes a point I made in my first lecture: a search into an applied problem can be challenging and exciting. But another lesson to take home is that nature is complex and we should not take a dogmatic stand when we face a biological problem. Let me quote the writer Paul Aldeston who said: "I have yet to see a problem however complicated that, when you look at it the right way, does not become more complicated." And finally we want to remember the simple admonition by Albert Einstein: "Everything should be made as simple as possible but not simpler."

Bibliography

Albers, R. W., Fahn, S., and Koval, G. J. (1963). *Proc. Natl. Acad. Sci. U.S.A.* **50,** 474.

Albers, R. W., Koval, G. J., and Siegel, G. J. (1968). *Mol. Pharmacol.* **4,** 324.

Albracht, S. P. J., and Slater, E. C. (1971). *Biochim. Biophys. Acta* **245,** 503.

Andreoli, T. E., Lam, K. W., and Sanadi, D. R. (1965). *J. Biol. Chem.* **240,** 2644.

Arion, W. J., and Wright, B. (1970). *Biochem. Biophys. Res. Commun.* **40,** 3.

Arion, W. J., and Racker, E. (1970). *J. Biol. Chem.* **245,** 5186.

Avron, M. (1963). *Biochim. Biophys. Acta* **77,** 699.

Avron, M., and Neumann, J. (1968). *Annu. Rev. Plant Physiol.* **19,** 137.

Avron, M., Grisario, V., and Sharon, N. (1965). *J. Biol. Chem.* **240,** 1381.

Avron, M., Bamberger, E. S., Rottenberg, H., and Schuldiner, S. (1973). *Proc. Int. Congr. Biochem., 9th, 1973* p. 215.

Azzi, A., Chance, B., Radda, G. K., and Lee, C. P. (1969). *Proc. Natl. Acad. Sci. U.S.A.* **62,** 612.

Bastide, F., Meissner, G., Fleischer, S., and Post, R. L. (1973). *J. Biol. Chem.* **248,** 8385.

Beechey, R. B., Holloway, C. T., Knight, I. G., and Robertson, A. M. (1966). *Biochem. Biophys. Res. Commun.* **23,** 75.

Bendall, D. S., and Hill, R. (1968). *Annu. Rev. Plant Physiol.* **19,** 167.

Bennun, A., and Racker, E. (1969). *J. Biol. Chem.* **244,** 1325.

Berden, J. A. (1972). Ph.D. Thesis, University of Amsterdam, Gerja, Waarland.

Bielawski, J., Thompson, T. E., and Lehninger, A. L. (1966). *Biochem. Biophys. Res. Commun.* **24,** 948.

Boguslavsky, L. I., Kondrashin, A. A., Kozlov, I. A., Metelsky, S. T., Skulachev, V. P., and Volkov, A. G. (1975). *FEBS Lett.* **50,** 223.

Boyer, P. D. (1958). *Proc. Int. Symp. Enzyme Chem. 1957*, p. 301.

Boyer, P. D. (1965). *In* "Oxidases and Related Redox Systems" (T. E. King, H. S. Mason, and M. Morrison, eds.), Vol. 2, p. 994. Wiley, New York.

Boyer, P. D. (1968). *In* "Biological Oxidation" (T. P. Singer, ed.), p. 193. Wiley, New York.

Boyer, P. D. (1974). *In* "Dynamics of Energy-Transducing Membranes" (L. Ernster, R. W. Estabrook, and E. C. Slater, eds.), p. 289. Elsevier, Amsterdam.

Bragg, P. D., Davies, P. L., and Hou, C. (1973). *Arch. Biochem. Biophys.* **159,** 664.

Bruni, A., and Racker, E. (1968). *J. Biol. Chem.* **243,** 962.

Bulos, B., and Racker, E. (1968a). *J. Biol. Chem.* **243,** 3891.

Bulos, B., and Racker, E. (1968b). *J. Biol. Chem.* **243,** 3901.

Burstein, C., Loyter, A., and Racker, E. (1971a). *J. Biol. Chem.* **246,** 4075.

Burstein, C., Kandrach, A., and Racker, E. (1971b). *J. Biol. Chem.* **246,** 4082.

Carmeli, C., and Avron, M. (1966). *Biochem. Biophys. Res. Commun.* **24,** 923.

Carmeli, C., and Racker, E. (1973). *J. Biol. Chem.* **248,** 8281.

Carmeli, C., Lifshitz, Y., and Gepshtein, A. (1975). *Biochim. Biophys. Acta* **376,** 249.

Caswell, A. H. (1971). *Arch. Biochem. Biophys.* **144,** 445.

Cattell, K. J., Knight, I. G., Lindop, C. R., and Beechey, R. B. (1970). *Biochem. J.* **125,** 169.

Catterall, W. A., and Pederson, P. L. (1971). *J. Biol. Chem.* **246,** 4987.

Chance, B. (1959). *Regul. Cell Metab. Ciba Found. Symp., 1958,* p. 91.

Chance, B. (1972a). *Mol. Basis Electron Transp. Proc. Miami Winter Symp. 1972,* Vol. **4,** p. 65.

Chance, B. (1972b). *FEBS Lett.* **23,** 3.

Chance, B., and Mela, L. (1967). *J. Biol. Chem.* **242,** 830.

Chance, B., and Schoener, B. (1966). *J. Biol. Chem.* **241,** 4567.

Chance, B., and Williams, G. R. (1956). *Adv. Enzymol.* **17,** 65.

Chien, T. F. (1974). *Fed. Proc., Fed. Am. Soc. Exp. Biol.* **33,** 1291.

Cockrell, R. S., and Racker, E. (1969). *Biochem. Biophys. Res. Commun.* **35,** 414.

Cockrell, R. S., Harris, E. J., and Pressman, B. C. (1967). *Nature (London)* **215,** 1487.

Cohn, M. (1953). *J. Biol. Chem.* **201,** 735.

Cohn, M. (1958). *J. Biol. Chem.* **230,** 369.

Cohn, M., and Drysdale, G. R. (1955). *J. Biol. Chem.* **216,** 831.

Cone, R. A. (1972). *Nature (London), New Biol.* **236,** 39.

Conover, T. E., Prairie, R. L., and Racker, E. (1963). *J. Biol. Chem.* **238,** 2831.

Cooper, C., and Lehninger, A. L. (1956). *J. Biol. Chem.* **219,** 489.

Danielli, J. F., and Davson, H. (1935). *J. Cell. Comp. Physiol.* **5,** 495.
Danon, A., and Stoeckenius, W. (1974). *Proc. Natl. Acad. Sci.. U.S.A.* **71,** 1234.
Davis, K. A., and Hatefi, Y. (1971). *Biochem. Biophys. Res. Commun.* **44,** 1338.
Deamer, D. W., and Baskin, R. J. (1972). *Arch. Biochem. Biophys.* **153,** 47.
Degani, C., and Boyer, P. D. (1973). *J. Biol. Chem.* **248,** 8222.
Deters, D. W., Racker, E., Nelson, N., and Nelson, H. (1975). *J. Biol. Chem.* **250,** 1041.
Eagle, H. (1971). *Science* **174,** 500.
Ebashi, S., and Lipmann, F. (1962). *J. Cell Biol.* **14,** 389.
Erecinska, M., Chance, B., Wilson, D. F., and Dutton, P. L. (1972). *Proc. Natl. Acad. Sci. U.S.A.* **69,** 50.
Ernster, C., and Lee, C. P. (1964). *Annu. Rev. Biochem.* **33,** 729.
Eytan, G., Carroll, R. C., Schatz, G., and Racker, E. (1975a). *J. Biol. Chem.* **250,** 8598.
Eytan, G., Matheson, M. J., and Racker, E. (1975b). *FEBS Lett.* **57,** 121.
Eytan, G., Schatz, G., and Racker, E. (1976). *Nobel Symp.* **34.**
Fahn, S., Hurley, M. R., Koval, G. J., and Albers, R. W. (1966). *J. Biol. Chem.* **241,** 1890.
Fahn, S., Koval, G. J., and Albers, R. W. (1968). *J. Biol. Chem.* **243,** 1993.
Farron, F. (1970). *Biochemistry* **9,** 3832.
Ferguson, S. J., Lloyd, W. J., Lyons, M. H., and Radda, G. K. (1975a). *Eur. J. Biochem.* **54,** 117.
Ferguson, S. J., Lloyd, W. J., and Radda, G. K. (1975b). *Eur. J. Biochem.* **54,** 127.
Fernández-Morán, H. (1962). *Circulation* **26,** 1039.
Fessenden, J. M., and Racker, E. (1966). *J. Biol. Chem.* **241,** 2483.
Fessenden-Raden, J. M. (1972a). *J. Biol. Chem.* **247,** 2351.
Fessenden-Raden, J. M. (1972b). *Biochem. Biophys. Res. Commun.* **46,** 1347.
Fessenden-Raden, J. M., Lange, A. J., Dannenberg, M. A., and Racker, E. (1969). *J. Biol. Chem.* **244,** 6656.
Fisher, R. J., Chen, J. C., Sani, B. P., Kapley, S. S., and Sanadi, D. R. (1971a), *Proc. Natl. Acad. Sci. U.S.A.* **68,** 2181.
Fisher, R. J., Sani, B. P., and Sanadi, D. R. (1971b). *Biochem. Biophys. Res. Commun.* **44,** 1394.
Fisher, R. J., Panet, R., Joshi, S., and Sanadi, D. R. (1973). *Biochem. Biophys. Res. Commun.* **54,** 469.
Folch, J., and Lees, M. (1951). *J. Biol. Chem.* **191,** 807.
Folch-Pi, J., and Stoffyn, P. J. (1972). *Ann. N. Y. Acad. Sci.* **195,** 86.
Forrest, G., and Edelstein, S. J. (1970). *J. Biol. Chem.* **245,** 6468.
Frye, C. D., and Edidin, M. (1970) *J. Cell Sci.* **7,** 319.

Futai, M., Sternweis, P. C., and Heppel, L. A., (1974). *Proc. Natl. Acad. Sci. U.S.A.* **71**, 2725.

Garber, M. P., and Steponkus, P. L. (1974). *J. Cell Biol.* **63**, 24.

Gasko, O. D., Knowles, A. F., Shertzer, H. G., Suolinna, E.-M., and Racker, E. (1976). *Anal. Biochem.* (in press).

Gatt, S., and Racker, E. (1959a). *J. Biol. Chem.* **234**, 1015.

Gatt, S., and Racker, E. (1959b). *J. Biol. Chem.* **234**, 1024.

Glynn, I. M., and Lew, V. L. (1969). *In* "Membrane Proteins," N. Y. Heart Assoc. Symp., p. 289. Little, Brown, Boston, Massachusetts.

Goldin, S. M., and Tong, S. W. (1974). *J. Biol. Chem.* **249**, 5907.

Goudard, M., and Chopin, J. (1974). *C. R. Hebd. Seances Acad. Sci., Ser. C.* **278**, 423.

Green, D. E., and Ji, S. (1972). *Mol. Basis Electron Transp. Proc. Miami Winter Symp., 1972* Vol. 4, p. 1.

Green, D. E., Lester, R. L., and Ziegler, D. M. (1956). *Biochim. Biophys. Acta* **21**, 80.

Gregory, P., and Racker, E. (1973). *Abstr., Int. Cong. Biochem., 9th, 1973* Sect. 4, Bioenergetics, 4L10, p. 238.

Groot, G. S. P., Kovac, L., and Schatz, G. (1971). *Proc. Natl. Acad. Sci. U.S.A.* **68**, 308.

Guillory, R. J., and Racker, E. (1968). *Biochim. Biophys. Acta* **153**, 490.

Gutman, M., Singer, T. P., and Beinert, H. (1971). *Biochem. Biophys. Res. Commun.* **44**, 1572.

Gutman, M., Singer, T. P., and Beinert, H. (1972). *Biochemistry* **11**, 556.

Haake, P. C., and Westheimer, F. H. (1961). *J. Am. Chem. Soc.* **83**, 1102.

Hackenbrock, C. R. (1966). *J. Cell Biol.* **30**, 269.

Harris, A. D., and Slater, E. C. (1975). *Biochim. Biophys. Acta* **387**, 335.

Hasselbach, W., and Makinose, M. (1962). *Biochem. Biophys. Res. Commun.* **7**, 132.

Hatefi, Y. (1968). *Proc. Natl. Acad. Sci. U.S.A.* **60**, 733.

Hatefi, Y., and Stempel, K. E. (1967). *Biochem. Biophys. Res. Commun.* **26**, 301.

Hatefi, Y., Haavik, A. G., Fowler, L. R., and Griffiths, D. E. (1962). *J. Biol. Chem.* **237**, 2661.

Hauska, G. A., McCarty, R. E., and Racker, E. (1970). *Biochim. Biophys. Acta* **197**, 206.

Hauska, G. A., McCarty, R. E., Berzborn, R., and Racker, E. (1971). *J. Biol. Chem.* **246**, 3524.

Higashiyama, T., Saunders, D. R., Serrianne, B. C., Steinmeier, R. C., and Wang, J. H. (1975). *Fed. Proc., Fed. Am. Soc. Exp. Biol.* **34**, 596 (abstr.).

Hilborn, D. A., and Hammes, G. G. (1973). *Biochemistry* **12**, 983.

Hilden, S., Rhee, H. M., and Hokin, L. E. (1974). *J. Biol. Chem.* **249**, 7432.

Hinkle, P., and Horstman, L. L. (1971). *J. Biol. Chem.* **246**, 6024.

Hinkle, P. C., and Leung, K. H. (1974). *In* "Membrane Proteins in Transport and Phosphorylation" (M. E. Klingenberg, E. Quagliariello, and N. Siliprandi, eds.), p. 73. North-Holland Publ. Amsterdam.
Hinkle, P. C., and Mitchell, P. (1970). *J. Bioenerg.* **1,** 45.
Hinkle, P. C., Penefsky, H. S., and Racker, E. (1967a). *J. Biol. Chem.* **242,** 1788.
Hinkle, P. C., Butow, R. A., Racker, E., and Chance, B. (1967b). *J. Biol. Chem.* **242,** 5169.
Hinkle, P. C., Kim, J.-J., and Racker, E. (1972). *J. Biol. Chem.* **247,** 1338.
Howell, S. H., and Moudrianakis, E. N. (1967). *Proc. Natl. Acad. Sci. U.S.A.* **58,** 1261.
Jagendorf, A. T. (1967). *Fed. Proc., Fed. Am. Soc. Exp. Biol.* **26,** 1361.
Jagendorf, A. T., and Neumann, J. (1965). *J. Biol. Chem.* **240,** 3210.
Jagendorf, A. T., and Smith, M. (1962). *Plant Physiol.* **37.** 135.
Jagendorf, A. T., and Uribe, E. (1966). *Proc. Natl. Acad. Sci. U.S.A.* **55,** 170.
Jardetzky, O. (1966). *Nature (London)* **211,** 969.
Johnson, L. W., Hughes, M. E., and Zilversmit, D. B. (1975). *Biochim. Biophys. Acta* **375,** 176.
Johnson, M. (1941). *Science* **94,** 200.
Kagawa, Y., and Racker, E. (1966a). *J. Biol. Chem.* **241,** 2461.
Kagawa, Y., and Racker, E. (1966b). *J. Biol. Chem.* **241,** 2467.
Kagawa, Y., and Racker, E. (1966c). *J. Biol. Chem.* **241,** 2475.
Kagawa, Y., and Racker, E. (1971). *J. Biol. Chem.* **246,** 5477.
Kagawa, Y., Kandrach, A., and Racker, E. (1973a). *J. Biol. Chem.* **248,** 676.
Kagawa, Y., Johnson, L. W., and Racker, E. (1973b). *Biochem. Biophys. Res. Commun.* **50,** 245.
Kanner, B. I., and Racker, E. (1975). *Biochem. Biophys. Res. Commun.* **64,** 1054.
Kanner, B. I., Serrano, R., Kandrach, A. M., and Racker, E. (1976). *Biochem. Biophys. Res. Comm.* (in press).
Katoh, S., and Takamiya, A. (1965). *Biochim. Biophys. Acta* **99,** 56.
Keilin, D. (1925). *Proc. R. Soc. London, Ser. B* **98,** 312.
Kielley, W. W., and Bronk, J. R. (1958). *J. Biol. Chem.* **230,** 521.
King, T. E. (1963). *J. Biol. Chem.* **238,** 4037.
King, T. E., Kuboyama, M., and Takemori, S. (1965). *In* "Oxidases and Related Redox Systems" (T. E. King, H. S. Mason, and M. Morrison, eds.), Vol. 2, p. 707. Wiley, New York.
Klingenberg, M. (1971). *In* "Energy Transduction in Respiration and Photosynthesis" (E. Quagliariello, S. Papa, and C. S. Rossi, eds.), p. 23. Adriatica Editrice, Bari.
Knowles, A. F., and Penefsky, H. S. (1972). *J. Biol. Chem.* **247,** 6617 and 6624.
Knowles, A. F., and Racker, E. (1975a). *J. Biol. Chem.* **250,** 1949.

Knowles, A. F., and Racker, E. (1975b). *J. Biol. Chem.* **250,** 3538.

Knowles, A. F., Guillory, R. J., and Racker, E. (1971). *J. Biol. Chem.* **246,** 2672.

Knowles, A. F., Kandrach, A., Racker, E., and Khorana, H. G. (1975). *J. Biol. Chem.* **250,** 1809.

Kornberg, R. D., and McConnell, H. M. (1971). *Proc. Natl. Acad. Sci. U.S.A.* **68,** 2564.

Krasne, S., Eisenman, G., and Szabo, G. (1971). *Science* **174,** 412.

Krebs, H. A. (1972). *Essays Biochem.* **8,** 1.

Krimsky, I., and Racker, E. (1955). *Science* **122,** 319.

Kyte, J. (1971). *J. Biol. Chem.* **246,** 4157.

LaBelle, E., and Racker, E. (1976). In preparation.

Lam, K. W., Warshaw, J. B., and Sanadi, D. R. (1967). *Arch. Biochem. Biophys.* **119,** 477.

Lambeth, D. O., and Lardy, H. A. (1971). *Eur. J. Biochem.* **22,** 355.

Lang, D. R., and Racker, E. (1974). *Biochim. Biophys. Acta* **333,** 180.

Lardy, H. A., and Elvehjem, C. A. (1945). *Annu. Rev. Biochem.* **14,** 1.

Lardy, H. A., and Parks, R. E., Jr. (1956). *In* "Enzymes: Units of Biological Structure and Function" (O. H. Gaebler, ed.), p. 584. Academic Press, New York.

Lardy, H. A., Johnson, D., and McMurray, W. C. (1958). *Arch. Biochem. Biophys.* **78,** 587.

Lee, C. P., and Ernster, L. (1966). *Regul. Metab. Processes Mitochondria, Proc. Symp., 1965* BBA Libr. Vol. 7, p. 218.

Lien, C.-T., and Racker, E. (1971). *In* "Methods in Enzymology," Vol. 23 (A. San Pietro, ed.), p. 547. Academic Press, New York.

Lien, S., and Racker, E. (1971). *J. Biol. Chem.* **246,** 4298.

Lindenmayer, G. E., Laughter, A. H., and Schwartz, A. (1968). *Arch. Biochem. Biophys.* **127,** 187.

Livne, A., and Racker, E. (1969). *J. Biol. Chem.* **244,** 1339.

Lonberg-Holm, K. K. (1959). *Biochim. Biophys. Acta* **35,** 464.

Luft, R., Ikkos, D., Palmieri, G., Ernster, L., and Afzelius, B. (1962). *J. Clin. Invest.* **41,** 1776.

Lynen, F. (1941). *Justus Liebigs Ann. Chem.* **546,** 120.

Lynen, F., and Konigsberger, R. (1951). *Justus Liebigs Ann. Chem.* **573,** 60.

Lynen, F., Hartman, G., Netter, K. F., and Schuegraf, A. (1959). *Regul. Cell Metab., Ciba Found. 1958* p. 256.

Lynn, W. S., and Straub, K. D. (1969). *Biochemistry* **8,** 4789.

McCarty, R. E., and Fagan, J. (1973). *Biochemistry* **12,** 1503.

McCarty, R. E., and Racker, E. (1966). *Brookhaven Symp. Biol.* **19,** 202.

McCarty, R. E., and Racker, E. (1967). *J. Biol. Chem.* **242,** 3435.

McCarty, R. E., and Racker, E. (1968). *J. Biol. Chem.* **243,** 129.

McCarty, R. E., Guillory, R. J., and Racker, E. (1965). *J. Biol. Chem.* **240,** PC4822.

McCauley, R., and Racker, E. (1973). *Mol. Cell. Biochem.* **1,** 73.
McCoy, G. D., and Racker, E. (1976). *Cancer Res.,* submitted.
McCoy, G. D., Resch, R., and Racker, E. (1976). *Cancer Res.,* submitted.
MacLennan, D. H. (1970). *J. Biol. Chem.* **245,** 4508.
MacLennan, D. H., and Tzagoloff, A. (1968). *Biochemistry* **7,** 1603.
MacLennan, D. H., Yip, C. C., Iles, G. H., and Seaman, P. (1972). *Cold Spring Harbor Symp. Quant. Biol.* **37,** 469.
MacMunn, C. A. (1914). "Spectrum Analysis Applied to Biology and Medicine." Longmans, Green, New York.
McMurray, W. C., Maley, G. F., and Lardy, H. A. (1958). *J. Biol. Chem.* **230,** 219.
Makinose, M. (1973). *FEBS Lett.* **37,** 140.
Makinose, M., and Hasselbach, W. (1971). *FEBS Lett.* **12,** 271.
Martonosi, A. (1967). *Biochem. Biophys. Res. Commun.* **29,** 753.
Martonosi, A. (1973). *Curr. Top. Membr. Transp.* **3,** 83.
Martonosi, A., Lagwinska, E., and Oliver, M. (1974). *Ann. N.Y. Acad. Sci.* **227,** 549.
Masuda, H., and de Meis, L. (1973). *Biochemistry* **12,** 4581.
Mitchell, P. (1961). *Nature (London)* **191,** 144.
Mitchell, P. (1966). *Biol. Rev. Cambridge Philos. Soc.* **41,** 445.
Mitchell, P. (1972). *FEBS Symp.* **28,** 353.
Mitchell, P. (1974). *FEBS Lett.* **43,** 189.
Mitchell, P. (1975). *FEBS Lett.* **56,** 1.
Mitchell, P., and Moyle, J. (1965). *Nature* **208,** 1205.
Mitchell, P., and Moyle, J. (1969). *Eur. J. Biochem.* **7,** 471.
Mitchell, R. A., Hill, R. D., and Boyer, P. D. (1967). *J. Biol. Chem.* **242,** 1793.
Moudrianakis, E. N. (1968). *Fed. Proc. Fed. Am. Soc. Exp. Biol.* **27,** 1180.
Mueller, P., and Rudin, D. O. (1969). *Curr. Top. Bioenerg.* **3,** 157.
Nakao, M., and Packer, L. (1973). "Organization of Energy-Transducing Membranes." Univ. of Tokyo Press, Tokyo.
Nelson, N., Nelson, H., and Racker, E. (1972a). *J. Biol. Chem.* **247,** 6506.
Nelson, N., Nelson, H., and Racker, E. (1972b). *J. Biol. Chem.* **247,** 7657.
Nelson, N., Deters, D. W., Nelson, H., and Racker, E. (1973). *J. Biol. Chem.* **248,** 2049.
Nelson, W. L., Butow, R. A., and Ciaccio, E. I. (1962). *Arch. Biochem. Biophys.* **96,** 500.
Nishibayashi-Yamashita, H., Cunningham, C., and Racker, E. (1972). *J. Biol. Chem.* **247,** 698.
Ochoa, S. (1943). *J. Biol. Chem.* **151,** 493.
Oesterhelt, D., and Stoeckenius, W. (1971). *Nature (London) New Biol.* **233,** 149.
Oesterhelt, D., and Stoeckenius, W. (1973). *Proc. Natl. Acad. Sci. U.S.A.* **70,** 2853.
Ogston, A. G., and Smithies, O. (1948). *Physiol. Rev.* **28,** 283.

Ohnishi, T. (1973). *Biochim. Biophys. Acta* **301,** 105.

Ohnishi, T., Asakura, T., Wohlrab, H., Yonetani, T., and Chance, B. (1970). *J. Biol. Chem.* **245,** 901.

Ohnishi, T., Wilson, D. F., Asakura, T., and Chance, B. (1972). *Biochem. Biophys. Res. Commun.* **46,** 1631.

Oleszko, S., and Moudrianakis, E. N. (1974). *J. Cell Biol.* **63,** 936.

Orme-Johnson, N. R., Orme-Johnson, W. H., Hansen, R. E., Beinert, H., and Hatefi, Y. (1971a). *Biochem. Biophys. Res. Commun.* **44,** 446.

Orme-Johnson, N. R., Hansen, R. E., and Beinert, H. (1971b). *Biochem. Biophys. Res. Commun.* **45,** 871.

Panet, R., and Selinger, Z. (1972). *Biochim. Biophys. Acta* **255,** 34.

Passonneau, J. V., and Lowry, O. H. (1962). *Biochem. Biophys. Res. Commun.* **7,** 10.

Pasteur, L. (1861). *C. R. Hebd. Seances Acad. Sci.* **52,** 1260.

Penefsky, H. S. (1967). *J. Biol. Chem.* **242,** 5789.

Penefsky, H. S., Pullman, M. E., Datta, A., and Racker, E. (1960). *J. Biol. Chem.* **235,** 3330.

Petrack, B., and Lipmann, F. (1961). *In* "Light and Life" (W. D. McElroy and H. B. Glass, eds.), p. 621. Johns Hopkins Press, Baltimore, Maryland.

Portis, A. R., and McCarty, R. E. (1974). *J. Biol. Chem.* **249,** 6250.

Post, R. L., Sen, A. K., and Rosenthal, A. S. (1965). *J. Biol. Chem.* **240,** 1437.

Post, R. L., Kume, S., and Rogers, F. N. (1973). *In* "Mechanisms in Bioenergetics" (G. F. Azzone *et al.* eds.), p. 203. Academic Press, New York.

Poyton, R. O., and Schatz, G. (1975). *J. Biol. Chem.* **250,** 762.

Pullman, M. E., and Monroy, G. C. (1963). *J. Biol. Chem.* **238,** 3762.

Pullman, M. E., Penefsky, H., and Racker, E. (1958). *Arch. Biochem. Biophys.* **76,** 227.

Pullman, M. E., Penefsky, H. S., Datta, A., and Racker, E. (1960). *J. Biol. Chem.* **235,** 3322.

Quastel, J. H. (1936). *In* "Perspectives in Biochemistry" (J. Needham and D. E. Green, eds.), p. 269. Cambridge Univ. Press, London and New York.

Racker, E. (1954). *Adv. Enzymol.* **15,** 141.

Racker, E. (1956). *Ann. N.Y. Acad. Sci.* **63,** 1017.

Racker, E. (1962). *Proc. Natl. Acad. Sci. U.S.A.* **48,** 1659.

Racker, E. (1963). *Biochem. Biophys. Res. Commun.* **10,** 435.

Racker, E. (1964). *Biochem. Biophys. Res. Commun.* **14,** 75.

Racker, E. (1965). "Mechanisms in Bioenergetics." Academic Press, New York.

Racker, E. (1967). *Fed. Proc., Fed. Am. Soc. Exp. Biol.* **26,** 1335.

Racker, E. (1970a). *Essays Biochem.* **6,** 1.

Racker, E. (1970b). *In* "Membranes of Mitochondria and Chloroplasts" (E. Racker, ed.), p. 127. Van Nostrand-Reinhold, Princeton, New Jersey.

Racker, E. (1972a). *J. Biol. Chem.* **247,** 8198.

Racker, E. (1972b). *J. Membr. Biol.* **10,** 221.

Racker, E. (1972c). *Am. Sci.* **60,** 56.

Racker, E. (1973). *Biochem. Biophys. Res. Commun.* **55,** 224.

Racker, E. (1974). *Mol. Cell. Biochem.* **5,** 17.

Racker, E. (1975). *In* "Energy Transducing Mechanisms" (E. Racker, ed.), p. 163. Butterworth, London.

Racker, E. (1975a). *Proc. FEBS Meet., 10th.,* **41,** 25.

Racker, E. (1975b). *In* "Proceedings of the International Symposium on Electron-Transfer Chains and Oxidative Phosphorylation" (E. Quagliariello *et al.,* eds.), p. 401. Fasano, North Holland Publ., Amsterdam.

Racker, E., and Eytan, E. (1973). *Biochem. Biophys. Res. Commun.* **55,** 174.

Racker, E., and Eytan, E. (1975). *J. Biol. Chem.* **250,** 7533.

Racker, E., Fessenden-Raden, J. M., Kandrach, M. A., Lam, K. W., and Sanadi, D. R. (1970). *Biochem. Biophys. Res. Comm.* **41,** 1474.

Racker, E., and Fisher, L. W. (1975). *Biochem. Biophys. Res. Commun.* **67,** 1144.

Racker, E., and Hinkle, P. C. (1974). *J. Membr. Biol.* **17,** 181.

Racker, E., and Horstman, L. L. (1967). *J. Biol. Chem.* **242,** 2547.

Racker, E., and Horstman, L. L. (1968). *Proc. Int. Congr. Biochem., 7th, 1967* Abstract II, p. 297.

Racker, E., and Horstman, L. L. (1972). *In* "Energy Metabolism and the Regulation of Metabolic Processes in Mitochondria" (M. A. Mehlman and R. W. Hanson, eds.), p. 1. Academic Press, New York.

Racker, E., and Kandrach, A. (1971). *J. Biol. Chem.* **246,** 7069.

Racker, E., and Kandrach, A. (1973). *J. Biol. Chem.* **248,** 5841.

Racker, E., and Krimsky, I. (1948). *J. Biol. Chem.* **173,** 519.

Racker, E., and Krimsky, I. (1952). *J. Biol. Chem.* **198,** 731.

Racker, E., and Stoeckenius, W. (1974). *J. Biol. Chem.* **249,** 662.

Racker, E., and Wu, R. (1959). *Regul. Cell Metab., Ciba Found. Symp. 1958,* p. 205.

Racker, E., Pullman, M. E., Penefsky, H. S., and Silverman, M. (1963). *Proc. Int. Congr. Biochem., 5th, 1961* Vol. V, p. 303.

Racker, E., Chance, B., and Parsons, D. F. (1964). *Fed. Proc., Fed. Am. Soc. Exp. Biol.* **23,** 431.

Racker, E., Tyler, D. D., Estabrook, R. W., Conover, T. E., Parsons, D. F., and Chance, B. (1965). *In* "Oxidases and Related Redox Systems" (T. E. King, H. S. Mason, and M. Morrison, eds.), p. 1077. Wiley, New York.

Racker, E., Horstman, L. L., Kling, D., and Fessenden-Raden, J. M. (1969). *J. Biol. Chem.* **244,** 6668.

Racker, E., Fessenden-Raden, J. M., Kandrach, M. A., Lam, K. W., and Sanadi, D. R. (1970a). *Biochem. Biophys. Res. Commun.* **41,** 1474.

Racker, E., Burstein, C., Loyter, A., and Christiansen, R. O. (1970b). *In* "Electron Transport and Energy Conservation" (J. M. Tager *et al.,* eds.), p. 235. Adriatica Editrice, Bari.

Racker, E., Hauska, G. A., Lien, S., Berzborn, R. J., and Nelson, N. (1972). *Proc. Int. Congr. Photosynth. Res., 2nd, 1971* p. 1097.

Racker, E., Knowles, A. F., and Eytan, E. (1975a). *Ann. N. Y. Acad. Sci.* **264,** 17.

Racker, E., Chien, T. F., and Kandrach, A. (1975b) *FEBS Lett.* **57,** 14.

Rafael, V. J., Klass, D., and Hohorst, H.-J. (1968). *Hoppe-Seyler's Z. Physiol. Chem.* **349,** 1711.

Ragan, C. I., and Garland, P. B. (1971). *Biochem. J.* **124,** 171.

Ragan, C. I., and Hinkle, P. C. (1975). *J. Biol. Chem.* **250,** 8472.

Ragan, C. I., and Racker, E. (1973a). *J. Biol. Chem.* **248,** 2563.

Ragan, C. I., and Racker, E. (1973b). *J. Biol. Chem.* **248,** 6876.

Rieske, J. S. (1971). *Arch. Biochem. Biophys.* **145,** 179.

Rieske, J. S., MacLennan, D. H., and Coleman, R. (1964). *Biochem. Biophys. Res. Commun.* **15,** 338.

Rose, I. A., Warms, J. V. B., and O'Connel, E. L. (1964). *Biochem. Biophys. Res. Commun.* **15,** 33.

Rottenberg, H., Grunwald, T., and Avron, M. (1972). *Eur. J. Biochem.* **25,** 54.

Rumberg, B., and Schröder, H. (1973). *Proc. Int. Photobiol. Congr. 6th, 1972* Abstract 036.

Ryrie, I., and Jagendorf, A. T. (1971). *J. Biol. Chem.* **246,** 3771.

Salach, J., Singer, T. P., and Bader, P. (1967). *J. Biol. Chem.* **242,** 4555.

Sanadi, D. R., Lam, K. W., and Ramakrishna Kurup, C. K. (1968). *Proc. Natl. Acad. Sci.* **61,** 277.

Schatz, G., and Mason, T. L. (1974). *Annu. Rev. Biochem.* **43,** 51.

Schatz, G., and Racker, E. (1966). *J. Biol. Chem.* **241,** 1429.

Schatz, G., Penefsky, H. S., and Racker, E. (1967). *J. Biol. Chem.* **242,** 2552.

Schatz, G., Groot, G. S. P., Mason, T., Rouslin, W., Wharton, D. C., and Saltzgaber, J. (1972). *Fed. Proc., Fed. Am. Soc. Exp. Biol.* **31,** 21.

Schneider, D. L., Kagawa, Y., and Racker, E. (1972). *J. Biol. Chem.* **247,** 4074.

Scholnick, P., Lang, D., and Racker, E. (1973). *J. Biol. Chem.* **248,** 5175.

Schuldiner, S., Rottenberg, H., and Avron, M. (1972). *Eur. J. Biochem.* **25,** 64.

Schwartz, J. H. (1963). *Proc. Natl. Acad. Sci. U.S.A.* **49,** 871.

Schwartz, J. H., Crestfield, A. M., and Lipmann, F. (1963). *Proc. Natl. Acad. Sci. U.S.A.* **49,** 722.

Selwyn, M. J., Dawson, A. P., Stockdale, M., and Gains, N. (1970). *Eur. J. Biochem.* **14,** 120.

Sen, A. K., Tobin, T., and Post. R. L. (1969). *J. Biol. Chem.* **244,** 6596.
Senior, A. E. (1973). *Biochim. Biophys. Acta* **301,** 249.
Senior, A. E., and Brooks, J. C. (1970). *Arch. Biochem. Biophys.* **140,** 257.
Serrano, R., Kanner, B., and Racker, E. (1976). *J. Biol. Chem.* (in press).
Shavit, N., Skye, G. E., and Boyer, P. D. (1967). *J. Biol. Chem.* **242,** 515.
Shertzer, H. G., and Racker, E. (1974). *J. Biol. Chem.* **249,** 1320.
Shinitzky, M., and Inbar, M. (1974). *J. Mol. Biol.* **85,** 603.
Siedow, J., Yocum, C. F., and SanPietro, A. (1973). *Curr. Top. Bioenerg.* **5,** 107.
Singer, S. J. and Nicolson, G. L. (1972) Science, **175,** 720.
Singer, T. P., and Gutman, M. (1970). *In* "Pyridine Nucleotide-dependent Dehydrogenases" (H. Sund, ed.), p. 375. Springer-Verlag, Berlin and New York.
Skulachev, V. P. (1971). *Curr. Top. Bioenerg.* **4,** 127.
Slater, E. C. (1953). *Nature (London)* **172,** 975.
Slater, E. C. (1972). *Mol. Basis Electron Transp, Proc. Miami Winter Symp. 1972,* Vol. 4, p. 95.
Slater, E. C. (1973). *Biochim. Biophys. Acta* **301,** 129.
Slater, E. C. (1974). *In* "Dynamics of Energy-Transducing Membranes," (L. Ernster, R. W. Estabrook, and E. C. Slater, eds.). Elsevier, Amsterdam. p. 1.
Slater, E. C., Lee, C. P., Berden, J. A., and Wegdam, H. J. (1970). *Biochim. Biophys. Acta* **223,** 354.
Smith, J. B., and Sternweis, P. C. (1975). *Biochem. Biophys. Res. Commun.* **62,** 764.
Smith, R. E., Roberts, J. C., and Hittelman, K. J. (1966). *Science* **154,** 653.
Sols, A. (1967). *In* "Aspects of Yeast Metabolism" (A. K. Mills and H. A. Krebs, eds.), p. 47. Blackwell, Oxford.
Stekhoven, F. S., Waitkus, R. F., and van Moerkerk, H. T. B. (1972). *Biochemistry* **11,** 1144.
Suolinna, E.-M., Lang, D., and Racker, E. (1974). *J. Natl. Cancer Inst.* **53,** 1515.
Suolinna, E.-M., Buchsbaum, R. N., and Racker, E. (1975). *Cancer Res.* **35,** 1865.
Taniguchi, K., and Post, R. L. (1975). *J. Biol. Chem.* **250,** 3010.
Telford, J. N., and Racker, E. (1973). *J. Cell Biol.* **57,** 580.
Thayer, W. S., and Hinkle, P. (1973a). *J. Biol. Chem.* **248,** 5395.
Thayer, W. S., and Hinkle, P. C. (1973b). *Fed. Proc. Fed. Am. Soc. Exp. Biol.* **32,** 669.
Thayer, W. S., and Hinkle, P. C. (1975). *J. Biol. Chem.* **250,** 5330 and 5342.
Tonomura, Y. (1972). "Muscle Proteins, Muscle Contraction and Cation Transport." Univ. of Tokyo Press, Tokyo.

Trebst, A. (1974). *Annu. Rev. Plant Physiol.* **25,** 423.
Tupper, J. T., and Tedeshi, H. (1969a). *Science* **166,** 1539.
Tupper, J. T., and Tedeshi, H. (1969b). *Proc. Natl. Acad. Sci. U.S.A.* **63,** 370.
Tupper, J. T., and Tedeshi, H. (1969c). *Proc. Natl. Acad. Sci. U.S.A.* **63,** 713.
Tyler, D. D. (1970). *Biochem. J.* **116,** 30.
Uyeda, K., and Racker, E. (1965a). *J. Biol. Chem.* **240,** 4682.
Uyeda, K., and Racker, E. (1965b). *J. Biol. Chem.* **240,** 4689.
Vambutas, V. K., and Racker, E. (1965). *J. Biol. Chem.* **240,** 2660.
Vanderkooi, G., Senior, A. E., Capaldi, R. A., and Hayashi, H. (1972). *Biochim. Biophys. Acta* **274,** 38.
Van De Stadt, R. J., De Boer, B. L., and Van Dam, K. (1973). *Biochim. Biophys. Acta* **292,** 338.
Wadkins, C. L., and Lehninger, A. L. (1963). *J. Biol. Chem.* **238,** 2555.
Walker, D. A., and Crofts, A. R. (1970). *Annu. Rev. Biochem.* **39,** 389.
Warburg, O. (1926). "Uber den Stoffwechsel der Tumoren." Springer-Verlag, Berlin and New York.
Warburg, O., and Christian, W. (1939). *Biochem. Z.* **303,** 40.
Warburg, O., Klotzsch, H., and Gawehn, K. (1954). *Z. Naturforsch. B.* **11,** 179.
Warren, G. B., Toon, D. A., Birdsall, N. J. M., Lee, A. G., and Metcalfe, J. C. (1974). *Proc. Natl. Acad. Sci. U.S.A.* **71,** 622.
Weinhouse, S. (1956). *Science* **124,** 267.
Whittam, R., and Ager, M. E. (1965). *Biochem. J.* **97,** 214.
Whittam, R., and Wheeler, K. P. (1970). *Annu. Rev. Physiol.* **32,** 21.
Wikström, M. K. F. (1973). *Biochim. Biophys. Acta* **301,** 155.
Wilson, D. F., and Dutton, P. L. (1970). *Arch. Biochem. Biophys.* **136,** 583.
Wollenweber, E., and Lebreton, P. (1971). *Biochimie* **53,** 935.
Wrigglesworth, J. M., Packer, L., and Branton, D. (1970). *Biochim. Biophys. Acta* **205,** 125.
Wu, R., and Racker, E. (1959a). *J. Biol. Chem.* **234,** 1029.
Wu, R., and Racker, E. (1959b). *J. Biol. Chem.* **234,** 1036.
Yamamoto, T., and Tonomura, Y. (1968). *J. Biochem. (Tokyo)* **64,** 137.
You, K., and Hatefi, Y. (1973). *Biochem. Biophys. Res. Commun.* **52,** 343.
Zalkin, H., Pullman, M. E., and Racker, E. (1965). *J. Biol. Chem.* **240,** 4011.
Ziegler, D., Lester, R., and Green, D. E. (1956). *Biochim. Biophys. Acta* **21,** 90.

Index

D

T

U

V

W

X

Y

A 6
B 7
C 8
D 9
E 0
F 1
G 2
H 3
I 4
J 5